Good Times
in the Badlands

Good Times
in the Badlands

Raleigh E. Emry

Writer's Showcase
San Jose New York Lincoln Shanghai

Good Times in the Badlands

Writer's Showcase
an imprint of iUniverse, Inc.

For information address:
iUniverse, Inc.
5220 S. 16th St., Suite 200
Lincoln, NE 68512
www.iuniverse.com

ISBN: 0-595-22673-6

Printed in the United States of America

Dedication

Photo 1: Morris F. Skinner—circa 1955

Dedicated
to the Memory
of
Morris F. Skinner
(1906-1989)

Contents

List of Illustrations

Foreword

Good Times in the Badlands embraces more than half a century of scientific fieldwork throughout the Great Plains of North America. It is about Morris F. Skinner, the leader of the expeditions, his wife Marie, who shared in the great adventure, and their cohorts who made the vast collections for the Frick Laboratory of the American Museum of Natural History in New York City.

If you took a poll of paleontologists, to discover what they believe is the most important or significant date, many might say, "sixty-five million years B.P." It was roughly sixty-five million years before present and near the close of the Cretaceous Period of the Mesozoic Era that certain events changed the course of life on earth. Worldwide volcanism may have caused reproductive stress in terrestrial animals. Acidification of the upper ocean could have caused equal stress on marine microorganisms. Perhaps the large meteorite striking the Earth near the Yucatan Peninsula also contributed to the demise of certain animal and plant life. The result was the abrupt extinction of dinosaurs and the rise of other species. Sixty-five million years B.P. marks the beginning of the Cenozoic Era, which is popularly known as the age of mammals. The familiar phrase "age of mammals" is accurate but restrictive. Birds, flowering plants and other major organisms also proliferated after that milestone date.

Morris Skinner, whose long career in paleontology focused on the Cenozoic Era, would have most surely chosen another date. His most important date was October 3, 1930 for it marks a milestone event that changed the course of Morris' life. It is the day that Shirley Marie White became his bride. October 3, 1930 was more important to Morris than his birth date as it was his new beginning. Morris never thought much about his legacy. Nonetheless, he did request that if a stone was erected in his memory, that October 3, 1930 be prominently carved upon it.

In 1973, Morris and Marie Skinner retired from their long careers. They had shared an association with the American Museum of Natural History that spanned their professional lives. In the winters, they conducted research at the museum; in the summers, they returned to their Ainsworth, Nebraska home for a season of fossil collecting throughout the western states. Upon their retirement, Marie was a Scientific Assistant in the Department of Vertebrate Paleontology and Morris was named Frick Curator Emeritus of the American Museum of Natural History.

This book about the Skinners and their associates is not a "how to" book, nor is it a scientific thesis. You can read about the science of paleontology in dozens of textbooks and dissertations. Expeditions to collect fossils with Morris were serious scientific endeavors and rife with physical hardships. Nonetheless, joyful enthusiasm and good humor filled each day. This book paints a picture of an era that is forever gone. It is freewheeling, poetic, humorous and nostalgic. Although fossil collecting is rooted in the *science* of paleontology, *Good Times in the Badlands* focuses on the effervescence or *spirit* of the profession instead.

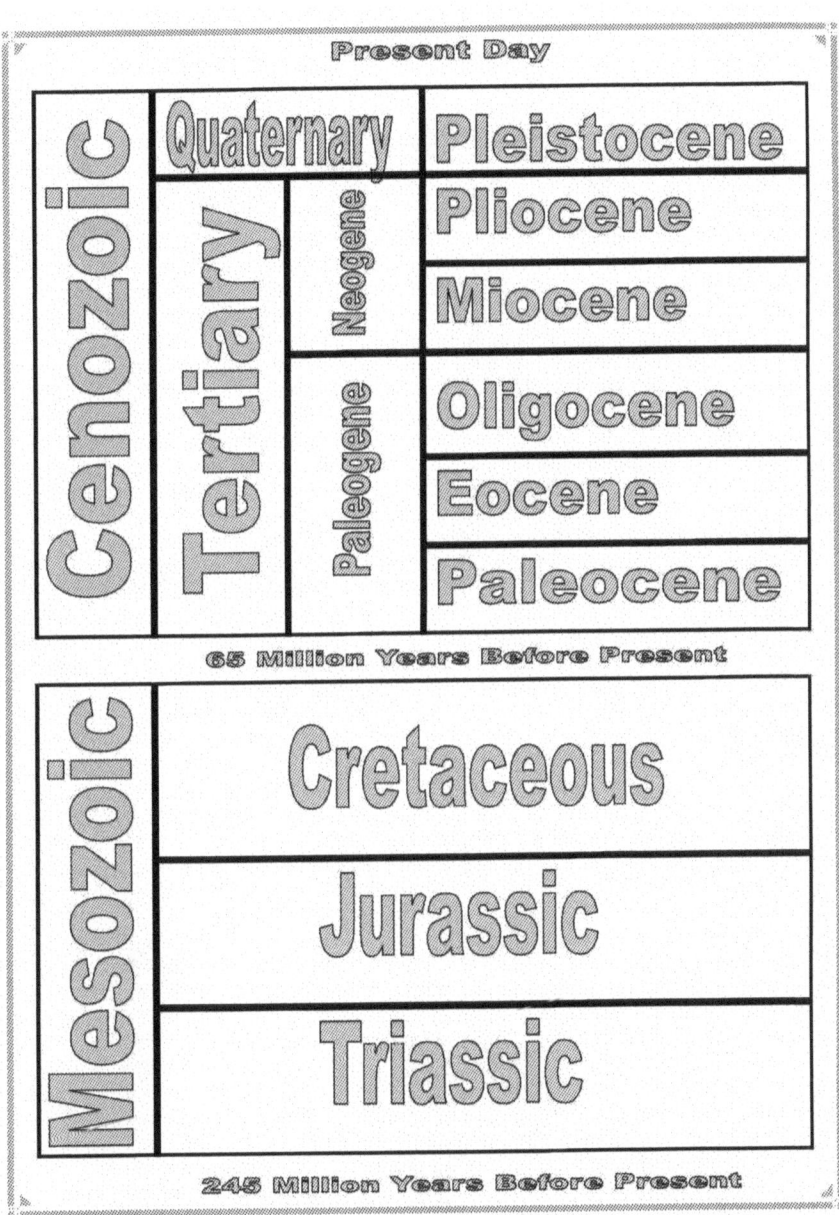

Figure 1: Major Divisions of Mesozoic and Cenozoic Eras

Acknowledgements

I want to extend my gratitude to all the people who I won't mention by name. Some of you offered anecdotes; some of you offered encouragement. Without encouragement, those anecdotes would be like fossil specimens that are still in their plaster jackets and hidden away in storage boxes. You motivated me to prepare them and place them on display within these pages. I also want to thank the many friends and acquaintances who read my manuscript and offered valuable advice.

Rod Worrell of the *Ainsworth Star-Journal*, Ainsworth, Nebraska, honored my request for photo negatives of Morris Skinner's old Model-A Ford pickup. Thanks Rod, for archiving seemingly insignificant things.

Judy Soles McMillie, a retired librarian of the *El Paso Times*, El Paso, Texas volunteered to go through my manuscript with a fine-toothed comb. Her contribution of time and talent immensely improved my work. I offer my sincere thanks, Judy, for tackling this difficult editing chore.

I also want to thank all of the fossil hunters who contributed memories of their days in the field with Morris Skinner. I hope I treated their words with the respect they deserve. Contributors include the late James H. Quinn, Tom Lucas, William J. Lear, the late John L. Beattie and Loren M. Toohey. Their essays are in chronological order throughout this book.

Loren M. Toohey, a retired EXXON petroleum geologist, came to my rescue whenever I needed scientific expertise. His interest in paleontology began in 1936 when he found fossils near his home in Hemingford,

Nebraska. Loren worked with the Nebraska State Museum field parties and then later with the Frick Laboratory of the American Museum of Natural History while attending the University of Nebraska and Princeton University. Loren still enjoys going to the field in search of fossils.

This book simply would not have existed without the interest, encouragement and direction of Morris Skinner's widow, Marie. She was not only the catalyst, but also the active ingredient, in bringing this project to completion.

In early 1998, and nearly 10 years after Morris' death, Marie invited the fossil hunters who had gone to the field with Morris to contribute their recollections. I am one such fossil hunter. Morris, Marie and I share the same hometown, Ainsworth, Nebraska. My summers collecting fossils with Morris encompassed those of my high school and college years from 1960 through 1966.

A few weeks after Marie's request, I sent her my draft entitled *Bone Digging Days*. Marie knew that I enjoyed writing and she saw something worthwhile in my essay. She was looking for someone to help her rewrite and organize the other essays that she had received. She also wanted someone to write about the early associates of Morris who are no longer with us. For those reasons, Marie asked me and I jumped in with both feet.

There is certainly a bit of Tom Sawyer in Marie Skinner. She invited me to whitewash a slat on her picket fence of bone digging memories. Once I had painted *my* picket, she lured me on down the line. A few times along the way, I was sure that I had run out of paint or that my brush was making a mess of it. Nevertheless, Marie would fill my bucket again and tell me that no one else could match my brush strokes. She enticed me along and made the project such fun that, before long, I had painted the whole fence. Thank you, Marie, for first trusting me with so much of your life story and then having such faith in my ability to put it on paper.

Finally, I would not have been able to begin this project without the love and support of my wife Terry. She teaches elementary school so that I can have the luxury of sitting at home and tinkering with words.

Introduction

Photo 2: From Sheep Mountain, South Dakota—July 1961

The Song of the Bone Digger
By Raleigh E. Emry

We travel light and easy
as we keep our *wants* at home
and only pack our *needs* to see us through.
When we are out with Morris Skinner,
heck, we barely need a comb.
A fossil hunter's *necessities* are so few.

A jug of alkali water
that's been warming in the sun
is hardly bottled Perrier or fancy stuff.
But it's nectar of the Gods when your tongue is dry
and rattlin' and your throat is parched
and voice is hoarse and rough.

You tip your jug across your arm
and swig it long and eager.
It floods out through your pores in instant sweat
that dries upon your hat brim
and your shirt in salty patterns.
If there's better brew, then I ain't found it yet.

The smell of bully beef a heatin'
and a pot of boiling taters,
makes your impatient belly grovel like a beast.
A can or two of peaches
to sate the sweet tooth's cravin's
rounds-out an appeasin' evening feast.

You attack the chow and sop the pot
with crusts of sun-hard bread
and swear you could eat another pot or two.
But the sunset comes a blazin'
and you're distracted by the colors
that range from crimson red to azure blue.

You wash up all the cook-pots
as the fiery panorama
fades slowly and so softly into night.
You yearn for that old bedroll
that's a bit rank, but so cozy.
You race the dark so you won't need a light.

You doze there like a baby
as the breezes stroke your body
and the Milky Way sparkles into view.
A nighthawk tips and teeters
and a coyote yodels lonely
a song that is so common, yet so new.

You swear it's only forty winks
when dawn finds the east horizon.
The coffee smell is heavy in the air.
Morris is up and at 'em.
The elusive fossil's waiting
for us to venture forth and find it there.

The coffee's thick as syrup
and as hot as molten lava.
A cup or two will wake up the deceased.
Some scrambled eggs and bacon
and some saturated hash browns
will help to keep your inner workings greased.

Your pack feels light and empty
as you swing it o'er your shoulder
and you take a drink to last you for the morn.
You hike into the badlands
with the air so fresh and sage-like.
It's days like this you're thankful you were born.

The early shadows highlight
the badland's scenic splendor.
You swear you must be walking on the moon.
But Ol' Sol keeps on heatin'
and as the shadows start retreatin'
you wish they wouldn't want to leave so soon.

Sunglasses are for sissies
and you've never heard of sun block.
Such things are for the lifeguards in the town.
So you squint into the brightness
with your sleeveless arms a browning
and tug the brim of your dusty Stetson down.

You've trained your eyes to focus on
the clues that few will notice.
A tiny glint of white against the gray
lures you ever closer
and you sit down to examine
the object that lies hidden in the clay.

Millions of years have come and gone
since this creature roamed the prairies.
You carefully brush the dust away from teeth.
They glisten like dark agate.
You smile and cross your fingers
and scratch the surface to reveal what lies beneath.

The earth and bones are ancient.
The sky and sun eternal.
Within all of this age-old vastness there's just you.
It gives you pause to wonder
why you have been the lucky chosen
to bring this fossil animal into view.

It's the song of the bone digger,
a relentless, curious chorus
that leads you ever-onward back in time.
It's the truth, clear and persistent,
calling from the rocks of ages,
a siren song so simple yet sublime.

PART 1
MORRIS, MARIE
AND THE
MUSEUM

Photo 3: Marie and Morris Skinner—Mid 1980s

Chapter 1

Morris F. Skinner

Morris F. Skinner (1906–1989) came from a family of pioneers who settled in north central Nebraska. His father, Fred, drove a freight wagon from the railroad at Ainsworth to outlying towns. Fred later owned a butcher shop, and then a hardware store in Ainsworth. During hunting season, he combined taxi and guide service for wealthy businessmen who came to fish and hunt in the Sandhill lakes.

Morris was strongly left-handed and had a reading disability, dyslexia, which was not often recognized back then. He repeated the third grade, trying to read letters and numbers as they were, instead of backward as he saw them. Confronting his disability, he doggedly pursued a reading course that lasted his entire life. In high school, he excelled in math, chemistry and physics, and all but failed in grammar and spelling. Morris was robust and adventuresome. He enjoyed playing football and baseball, and shinnied up telephone poles or windmills just to teeter on the heights.

After Morris graduated from high school in 1925, General Electric of Fort Wayne, Indiana, accepted him for a year of intensive machine shop practice. Morris took his training to the oil town of Borger, Texas where he applied his new trade. He returned to Ainsworth in 1927. Once back home, he found a job as night attendant for a hydroelectric plant on Plum Creek, a tributary of the Niobrara River, northwest of Ainsworth.

3

Morris believed that the disciplines that he learned during those years were to be as important as his later college courses. For example, his practical training resulted in sixty years of sharpening and tempering field picks and laboratory tools. When he needed to measure vertical sections of the fossil beds, he made a hand-level from a piece of brass. It is now in the American Museum of Natural History along with his section books and diaries where he recorded his findings. During World War II when Great Britain could not furnish awls, Morris made them from old bedsprings.

It was in the Plum Creek canyons near the hydroelectric plant that Morris began to find fossils and he sold them to finance his college education at the University of Nebraska. In 1932, he graduated with a Bachelor of Science Degree in Geology. In 1933, Mr. Childs Frick invited Morris to the American Museum of Natural History in New York City to prepare and study fossils there in his Frick Laboratory.

You cannot appreciate this chronology of dates without my first putting them into a more global context. When I remind you of the seriousness of concurrent world events, you will see that this 1933 coincidence between Childs Frick and Morris Skinner was as extraordinary as the rare conjunction of two celestial bodies.

On October 24, 1929, a day that is forever known as "Black Thursday," the stock market crashed. Some investors, realizing that their fortunes were now as worthless as the ticker tape that forecast their doom, leaped from their office windows to their deaths. Black Thursday ushered in more than a decade of the most terrible economic times this country has ever known. This worldwide economic catastrophe was named the "Great Depression" to differentiate it from all lesser ones. Large portions of the unemployed population were now standing in bread lines, when days before they considered their futures bright.

Farmers in the Great Plains were not spared. Increased crop production in the '20s had an adverse effect of depressing prices of their products. A record wheat crop in 1931 drove prices even lower and there was now little money to pay for their produce. Although farmers were more productive than ever before, the fruits of their labor would no longer pay for their expenses. Then, with a vengeance, came years of drought—dry, windy, depressing days that turned this breadbasket of the nation, the Great Plains, into a dust bowl. By 1933, farmers all across the land were voluntarily deeding their farms to creditors, losing their land to bank foreclosures, or leaving farms temporarily in a near hopeless quest to provide for their families.

In 1930, the year after the stock market crash, optimists felt the worst was over. "Happy Days are Here Again" was the song of the year even though the national economy shrank from 87 billion to 75 billion dollars. "I've Got Five Dollars" was the theme song of 1931 as the economy retreated even further. By 1932, when the economy precariously hovered at only a third of its 1929 value, people were singing the mournful lament "Brother Can You Spare a Dime?" and they sang it for the rest of the '30s.[1]

In spite of this unprecedented national turmoil, two people seemed all but oblivious to it. A millionaire, Childs Frick, in New York City had preserved sufficient capital to continue his keen interest in prehistoric life and to pay for fossil hunting expeditions. A young man in the Nebraska dustbowl, Morris Skinner, wanted nothing else. The pay was meager but Morris considered himself fortunate. When great cities were dealing with lines of beggars and people willingly worked for pennies, this merging of interests between these two men motivated them beyond low pay, depressions, droughts, politics or bread lines. This coalition endured until Mr. Frick's death in 1965.

As the Frick Laboratory specialized in mammalian fossils, the expeditions in search of fossils were predominately to sediments of the Medial and Late Cenozoic Era. The sediments exposed in canyons near

---

[1] *American History 102, Civil War to the Present,* University of Wisconsin, Stanley K. Schultz, Ph.D.

Morris' home in Nebraska are of this age. Shared interests combined with a convenience of geology helped facilitate the bond between Frick and Skinner.

Although mammals were the focus of the Frick expeditions, many types of plants and animals existed in prehistoric times, just as they do today. The Frick Collection therefore includes fossil birds, reptiles, mollusks, plant seeds, petrified wood, etc. Species not included in the collection are the creatures that most laymen might think of first, the dinosaurs. Dinosaurs existed at an earlier prehistoric time.

When the collection season was over each year, Morris moved his family to New York City. From the 1930s to the mid-50s, they lived in brownstone, furnished rooms close to Public School, PS 87, and a liberal Protestant church on 76th and Central Park West. Their children, Barbara and Fritz, had friends in PS 87 and played in the church many afternoons. The church had classes in art and music and the supervisors there took the children to the Metropolitan Museum and Central Park.

When Morris and Marie found an apartment they could afford within walking distance of work, they promptly furnished it from Tepper's Gallery, an auction house that adjoined the apartment house. They enjoyed this place and so did others who worked at the museum when Morris and Marie were in the field.

During his months at the museum, Morris put his organizational skills to use. With the cooperation of Floyd Blair, Registrar of the Frick Laboratory, Morris worked out a system of field lists that served as basic catalogues. He then supplied all of the supporting geographic and geologic data. This system remains in use for all collections of the Frick Laboratory.

Morris grouped and organized antelope, deer, peccary, rhinoceros, bison and horse specimens. With the arrival of spring each year, Morris and his family were itching to get back to the field. The emphasis of this book is the fieldwork that Morris conducted for over half a century throughout the Great Plains.

The Good Neighbor

During the Great Depression, Morris often referred to his employment as "glorified relief." "Relief" was the word for "welfare" then. Like J.P. Morgan, Andrew Carnegie and many others, Childs Frick's father, Henry Clay Frick, was an entrepreneur of the late 19th century. These American capitalists accepted the many risks associated with the acquisition of great wealth. Exploitation and ruthlessness accompanied this acquisition. Henry's son, Childs, was a philanthropist. He donated his fortunes to many worthy endeavors including scientific research. His passion was mammalian fossils. He sent field parties far and wide around the world.

Even a millionaire's fortunes can be depleted. Therefore, Mr. Frick funded the research, but did so frugally. Morris' salary was nearly in line with the disadvantaged workers in the steel mills and coke kilns of the past generation. Nonetheless, Morris had a job and he shared his pay with needy folks back home.

Morris hired impoverished farmers and their teams of horses, by the day, to remove overburden from his fossil quarries. The pay for a day's work was only two or three dollars. Farmers who were determined to hold onto their land during years of drought deeply appreciated their being able to trade hard work for needed cash.

Agatha Anderson Hall, a farmer's daughter, recalls such an occasion in 1934 when she was a young girl. Morris hired her father, George Anderson, and his team of horses for three dollars per day. His team pulled a cable attached to a scraper in a renowned fossil locality, Devil's Gulch, northeast of Ainsworth and near the Anderson farm. Agatha remembers how fortunate her father felt about getting work. The three dollars each day meant survival. Agatha also remembers her excitement when her father told her that Morris had found a three-toed horse!

Showing the rural folks the incredible things that were buried in the gravels, sands, and clays beneath their cornrows or pastures of prairie grass enhanced the camaraderie between Morris and the landowners. A three-toed horse was not uncommon in the fossil beds of Northern Nebraska. Nevertheless, when a farm family had seen the evidence that such creatures existed millions of years before, even the ordinary scientific discoveries brought a new dimension to their rural life.

As Morris traveled throughout the Niobrara canyon land, he often saw solutions to some of the farmer's dilemmas. He noticed that Bill Clark at Meadville had hungry pigs and that Hans Johnson, who farmed upriver near Sparks, had a supply of corn but with nothing to eat it. Morris loaded his pickup with some of Bill Clark's pigs and drove them to Hans Johnson's farm. He then returned a pickup load of corn to Bill Clark. Both farmers considered it an even trade.

Morris always stopped at the farmhouse to chat before visits to the canyons. He usually stopped again on the way out for a drink of water, to tell them what he had found, and to give them an assessment of their livestock that he may have seen in distant pastures. Morris always made sure that gates were left in the position he found them. He also carried a bucket of fencing staples to tack up the barbwire on fences in disrepair. When notified of a prairie fire, Morris and his crew would rush to the scene and work until the fire was under control. Through these simple courtesies, Morris gained immense respect. The fossilized bones eroding from the canyon walls were fair payment for Morris' being a good neighbor and sharing the Frick fortune a dollar at a time.

Making Sense of It All

Fossil hunting expeditions are of generally two types—prospecting and quarry work.

The hunt for fossils always begins with prospecting, the simple act of walking the canyons or badlands to study the geology and to look for the fossilized remains of animals. Isolated fossils are throughout the sediments just as skeletons of modern-day animals are scattered across the prairies. Most collecting areas in the Great Plains, such as the Big Badlands of South Dakota, are a prospector's domain. Prospecting then is a process of being on the move, sleeping under the stars in bedrolls, and collecting the isolated fossils that you find.

If the search was fruitful, Morris assigned a "Prospecting Locality" name to the area and he listed and numbered the fossils found there accordingly. Prospecting Localities are along streams and canyons that cut into the fossil producing layers and in areas of exposed sediments around mesas, buttes and badlands. These exposed sediments are often called outcrops. A Prospecting Locality along a stream could be as narrow as the tributary for which it was named, and as long as five or six miles such as the "Snake River Locality" or "Deep Creek Locality" in Nebraska. Conversely, the "McGill Ranch Locality" was an area of outcrops that extended throughout 160 acres, or a square of badlands with one-half mile sides.

The Prospecting Locality always described the geographic location. Numbers and a description assigned to each specimen pinpointed the location of the find and tied the fossil to the stratigraphy or the geology. This precise documentation is extremely important; if a specimen is not linked to the geology, it has little scientific value. The Prospecting Locality numbers ensured that isolated specimens from 850 boxes of fossils, from 515 different localities, could be catalogued, cross-referenced, and located.

Prospecting Localities produce a broad range of animal types from tiny moles or shrews to large land mammals such as mastodon and titanothere. The ages of mammalian specimens throughout a Prospecting Locality could possibly span sixty-five million years. The relative ages could range from the Dawn of Mammals, in the Early Cenozoic sediments, to Ice Age and recent mammals, in the uppermost layers of the same butte or mesa.

If Morris found more than a dozen fossils in close proximity, he then gave that specific place a quarry name. Some quarries are extensive and will still produce fossils. Others were isolated channel deposits that were completely excavated.

Fossil quarries are how the public might picture fossil collecting. Dinosaur National Monument in Utah, or Agate Fossil Beds National Monument and Ashfall State Historic Park in Nebraska, are examples of a fossil quarry.

A jumble of bones in close proximity is the result of some concentrating phenomenon near the time when the animals perished. It could have been a water hole where animals gathered, or it might have been a channel deposit where ancient streams, perhaps floods, swept up animals and deposited their bones in the way that logjams are formed.

Ashfall State Historic Park in Nebraska, a fossil quarry discovered by Mike Voorhies of the University of Nebraska State Museum, is currently being excavated under his direction and can be observed by the public. The rhinos and other fossils at Ashfall are buried in volcanic ash and are a snapshot of an event of short duration when a volcano erupted and buried them. Quarries therefore produce animals that shared a specific time and place. They are valuable in establishing the fauna of that precise prehistoric moment.

Therefore, when you read "prospecting," imagine roaming over large expanses of badlands with a pack on your back. When you read "quarry," imagine first the backbreaking labor needed to expose the isolated deposits of fossilized bones.

In addition to collecting fossils, Morris made sections of the stratigraphy. This meant that he measured and described the exposed sedimentary layers in the fossil localities. The geology that contains fossil land animals usually consists of layers of wind-born and water-born deposits. Some of these layers are volcanic ash.

In about 1960, Morris and his field crew began to collect hundreds of volcanic ash samples from throughout the prospecting localities. This emphasis was due to some new discoveries at the University of California's Department of Geology. Garniss H. Curtis and his partner Jack F. Evernden had devised an "atomic clock" that measures the decay of potassium 40 into argon 40. This rate of change is enormously slow, but it is constant. The amount of change leads to discovering how long ago it started. As volcanic tuff or ash contains potassium-bearing minerals, the "atomic clock" could assign dates to the times of the volcanic eruptions that produced the ash.[2]

In the 1980s, the isotopes of argon 40/argon 39 replaced the potassium-argon method; it is a method with less contamination and fewer errors. This constantly evolving science of radiometric dating of the volcanic eruption then sets accurate ages for the fossils found within the ash, or a range of ages for the fossils found between volcanic ash layers. For the more recent specimens from the late Pleistocene forward, the radioactive decay of carbon 14 is used.

Laboratories have since analyzed some of the samples that Morris collected and they have provided us with a chronology for different events. In summary, fossil collecting is a synchronized effort to tie the specimen to its proper level in a column of sediments that could be thought of as a tower of time.

---

[2] *National Geographic*, Oct. 1961, Pg. 590-592, "A Clock for the Ages: Potassium-Argon", Garniss H. Curtis, Ph.D.

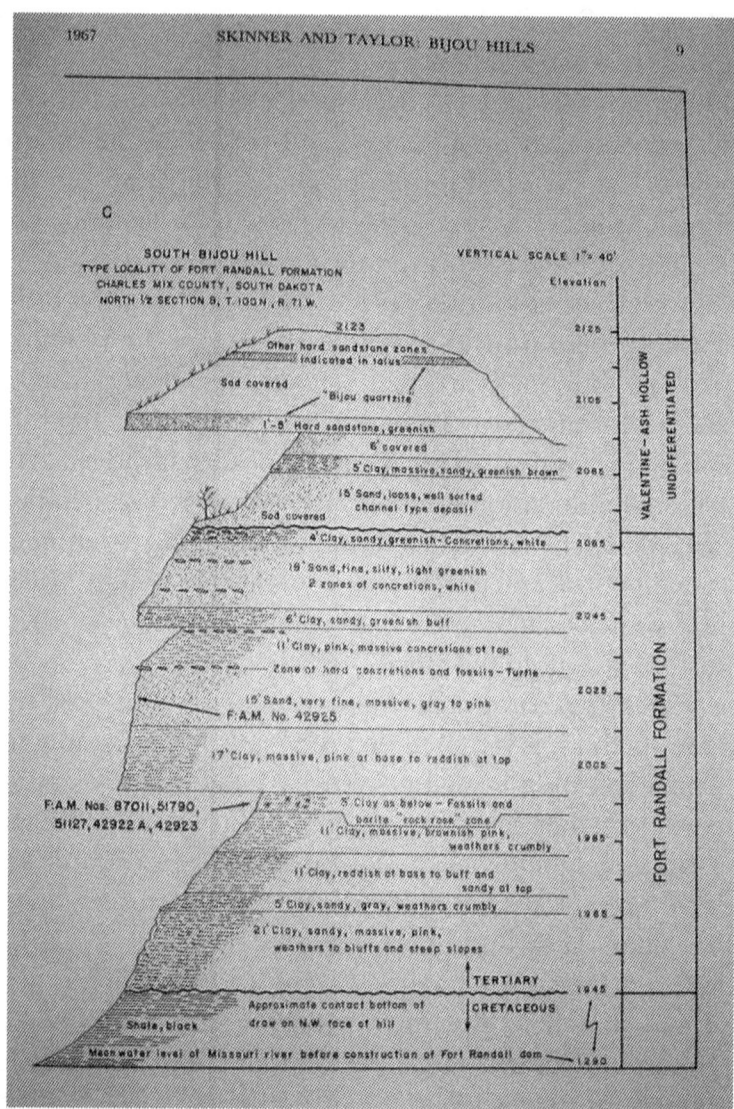

Figure 2: Example of a Geologic Section.

---

[3] A Revision of the Geology and Paleontology of the Bijou Hills, South Dakota, Pg. 9, Morris F. Skinner and Beryl E. Taylor, *American Museum Novitates,* No 2300, Sept 15, 1967

Photo 4: Maudie—Morris Skinner's 1928 Model-A Ford Pickup

The Men and Maudie

Fieldwork requires assistants. Summers of prospecting might have found Morris in the field with only one or two assistants. Summers of quarry work required more manpower.

In the early years, some of Morris' assistants came to the field from the museum in New York City. They were like fish out of water. The museum men enjoyed a summer outing, but found the quarry work too hard and the prospecting, with the idea of coming upon a rattlesnake, too unsettling. They saw the narrow ribbons of prairie highways, which led to the summer headquarters in Ainsworth, Nebraska, as more dangerous than side streets on Long Island. Their assessment of food in camp was dismal, the sleeping conditions were worse, and small towns offered little for their entertainment.

With little to lose, Morris decided to hire responsible, local men. The men, usually teenagers, prospected and worked in the quarries with boundless energy and enthusiasm. They were accustomed to the rigors of rural life and hard work. Working for Morris was hard work indeed, but it was interesting, very satisfying, hard work. The local boys were agile, interested, competitive and energetic and made a game of finding rocks and fossils. Their high spirits and laughter were good indications of their love for their work.

Some men who collected fossils with Morris continued lifelong careers in geology or paleontology. Others struck off on other routes through life. For example, I completed a career in the military as an Air Force officer. My brother, Robert Emry also started out as an assistant to Morris but stayed with it. Brother Bob has enjoyed a long career as Curator of Vertebrate Paleontology at the Smithsonian Institution's National Museum of Natural History specializing in the evolutionary history of Cenozoic mammals in the Northern Hemisphere.

Although the men in this book are from diverse backgrounds, there is a common thread that weaves their stories together. When you read their words, you will recognize this obvious, yet subtle, coherence and understand that it is easier to witness than to describe. It may be the love of the natural world and the thrill of new discoveries. It may be about friendship and respect between folks who share a common interest. It may simply be nostalgia for exciting times long ago in the company of Morris and Marie

One common thread that most fossil collectors under Morris' supervision share is their preference for prospecting over quarry work. Quarry work is extremely labor intensive and reminded some of the local assistants of the "quarry work" that they might have already done, mining their parents' cattle barns or horse stables. The fun of quarry work, removing the fossils from their resting places with whiskbrooms and small hand tools, only begins after the backbreaking drudgery of removing the layers above the fossils, the overburden, with picks and shovels.

If the overburden was extensive, Morris sometimes brought in machines or horse-drawn scrapers to assist. In the early days of quarry work with Morris, the nation was suffering a depression, a drought, and then a World War. Machines were hard to come by. Therefore, Morris improvised by modifying a 1928 Model-A Ford pickup with an extra transmission to produce sufficient power to not only take them on off-road prospecting expeditions, but to be used as a tractor in the quarries as well.

His contraption was loved by all who knew her. She took on a personality and soon a series of names. "Old Maudie" or simply "Maudie" is how she is remembered. Few spoke of her in disparaging ways, as they knew she could do the work of ten men. Therefore, when she broke down, someone might affectionately say, "Maudie cast her withers again." This obsolete statement describes a workhorse that had suffered a shoulder problem and could no longer pull.

Those who knew Maudie mention her time and again and always in anthropomorphic ways. Observers, who were not aware of Maudie's modifications, were always surprised to see a stripped-down, Model-A Ford pickup doing things a Model-A Ford could not do. Morris sometimes amused the newcomers by whipping his "Iron Mule" playfully with a switch and goading her up seemingly vertical landscapes.

Those who didn't respect Maudie were quickly taught to mind their manners. Two young assistants, during their first summer in the field, decided to make their mark in bone digging by branding Maudie with their initials. Morris frowned at their handiwork and told them in words they would not soon forget, that better bone diggers than they had managed not to brand Maudie during their many years of employment.

Maudie was so necessary to those early fossil hunting expeditions that, some fifty years after her retirement, University of Nebraska pale-ontologist, Mike Voorhies, in a statement of profound respect for both Maudie and Morris, proclaimed her "a scientific instrument invented and used by the best of them all."[4]

In 1995, Maudie was taken out of retirement from a shed on Marie Skinner's Ainsworth, Nebraska farm and was shipped by truck to New York City. The American Museum of Natural History commissioned the Horseless Carriage Carriers, Inc., of Paterson, New Jersey, to haul her there and back. The van was an impressive conveyance that usually carried polished specimens like vintage Cadillacs or Pierce Arrows. Nonetheless, Maudie looked completely at ease, a work brittle, gallant, old lady going off on another bone digging expedition.

The $3000 dollar round trip fee is an indication of how important Maudie was to the museum collections. On October 7, 1995, she opened in New York City as a centerpiece in an exhibit called "A Century of Discoveries: Fossil Expeditions of the American Museum of Natural History".

To understand Maudie is to understand Morris. Both were tough, unvarnished, determined, sometimes cantankerous adventurers, but their associates admired and respected them. Maudie's contribution to the 20th Century was not forgotten, nor did the world forget Morris. At the close of the Second Millennium, Morris F. Skinner was chosen as one of the "100 Extraordinary People of Nebraska."[5]

---

[4] *Ainsworth Star Journal*, Ainsworth, Nebraska 69210, Front Page, Vol. 114, No. 43, 25 Oct 1995.

[5] *Omaha World Herald*, "Celebrating A Century, 100 Extraordinary People of Nebraska," Omaha, Nebraska, November 28, 1999.

Photo 5: Morris collecting a volcanic ash sample near Chimney Rock, Nebraska—Oct 1961

In Search of the Truth

One of Morris' extraordinary traits was his dogged pursuit of the truth. For as Morris always said, "if you don't have the correct geology to go with your discoveries, all you have are some pretty rocks."

If that meant correcting other published interpretations, Morris was quick to do so. He studied the diaries and journals of early explorers and sometimes found conflicts between their writings and later literature. For example, two fossil hunters and explorers, F.V. Hayden and F.B. Meek, had visited the upper Missouri River in 1853. Morris became suspicious that the Bijou Hills, as listed on the maps of South Dakota, were not the hills that Hayden had described and named.

The Bijou Hills were inaccessible by automobile. Therefore, in 1937, Morris set out on the Missouri River by boat, as did the Hayden expedition, and found the true Bijou Hills exactly as Hayden had described them, as prominent landmarks along the Missouri River.[6]

His discovery was disconcerting to the South Dakota Geological Survey, as they had mistakenly placed the earlier named hills, both geographically as well as geologically, in a different place. A miffed South Dakota geologist told Morris that he should stay in his own state. Morris viewed his employment with the Frick Laboratory as "national" in scope and not tied to a specific state. Therefore, Morris corrected the records wherever he went, whether or not his efforts were appreciated.

Sometimes his corrections were as subtle as sending a colleague a penny postcard of Chimney Rock. The colleague had stated, at an earlier time, that the volcanic ash layers they were discussing were not present on Chimney Rock. The photo of the Oregon Trail landmark clearly showed the volcanic ash layers. Morris boldly annotated the ash layers and penned a wry message that went something like this, "Having a wonderful time. Wish you were here. You ought to visit Chimney Rock sometime."

Morris had little respect for the professors, paleontologists or geologists who published lengthy dissertations without spending much time in the field. Morris would often join such folks on their fieldtrips of "road-cut geology." By stopping at road-cuts, you can study the geology exposed by heavy construction machinery without venturing more than a hundred yards from your automobiles. There is nothing wrong with such an adventure. It has its benefits. Organized road-cut tours

---

[6] A Revision of the Geology and Paleontology of the Bijou Hills, South Dakota, by Morris F. Skinner and Beryl E. Taylor, *American Museum Novitates*, No 2300, Sept 15, 1967

offer good evidence and a caravan of like-minded folks can have a very good time.

Morris' perpetual field uniform was a battered felt Stetson or western style straw hat, tan work pants and shirt, and a pair of scuffed, durable work shoes that seemed more suited for a farm than for a badland safari. If Morris thought an "authority" with nary a scuff on his expensive hiking boots had not ventured far beyond road-cuts, Morris would offer up his trademark declaration. With a half-smile, half-sneer and squinted eyes unprotected from the bright sun—a face formed that way by decades of studying geology in the most remote corners of the map—Morris would remark, "Boys! You have a lot to learn!"

Understated Expertise

Morris was always low-key about his accomplishments. He did things for the doing and was often chagrined when his work brought admiring attention to himself. He sometimes joked about the pleasant greetings he would receive from passersby on the streets of Ainsworth. His account of the conversation was often in high falsetto to mimic the greeter, "Oh my! Doctor *Skinner*, your work is so *interesting*!"

He began his autobiography on September 11, 1986, when he was eighty years old and summed up his life in these few paragraphs.

"I was born in Springview, Nebraska on September 14, 1906, and grew up in a typical small town, Ainsworth, Nebraska. I attended grade school and High School and graduated in 1925. During my early school years, I became well versed in the ways of outdoor life and the ways of nature as it was found in Nebraska.

In school, I became very interested in history, electronics and radio and read everything I could about these subjects. One of my first "jobs" was helping install telephones, making radios and doing home wiring.

After finishing high school, I went to Fort Wayne, Indiana where I had heard of the General Electric Apprentice School. I applied and took a competitive examination and was accepted into a small group of apprentices that General Electric trained each year. We were trained in general shop practices, such as the operation of lathes, milling, machines, recognition of tools and quality of tool steels and a general course in metallurgy.

The course lasted for nine months after which I returned to Ainsworth.

I heard of a new oil discovery near Borger, Texas and I went. I had the experience of working in a "Boom Town" and put my General Electric training to use in a "Phillips Oil" machine shop for a winter season.

I returned to Ainsworth and obtained a position as night man for the Ainsworth Hydroelectric Plant. As a night man, my early interest in things natural took over and I found plenty of time during the day to look for "pretty rocks" in the canyons of Plum Creek where the hydro plant was located.

A kindred spirit in the person of a farm boy, Jim Quinn, was my buddy. It was only natural that we should meet and become lifelong friends. It was not long until we had become interested in the rocks and fossil remains that we found in the canyons. We could only guess at the fossils and the kinds of animals that existed and produced the bones and teeth we found. One particular specimen was the horizontal ramus of a Rhinoceros with well-preserved teeth and bone.

We sent the rhino jaw to E. H. Barbour PhD, at the University of Nebraska State Museum, who identified our find. He wrote back and said he could give us $10 for the jaw and if we could find more of the animal, it would be worth much more scientifically.

Dr. Barbour, no doubt, had only one thought, that of encouragement. Fossils were known to exist in Brown County, Nebraska because of A. C. Whitford's find in Devil's Gulch. Little did Dr. Barbour know that Jim and I knew where there was a solid mass of bones from which we obtained the original jaw!

During the rest of the fall of 1926, Jim and I collected bones from the deposit we had blundered on. We used flour and water paste to cast a sizable number of specimens that included at least two skulls, several jaws and numerous ribs, vertebrae and limbs. We sent these off to the museum and sat back waiting to become rich.

Dr. Barbour replied that he could give us $125 for the whole collection. With that letter, we became aware that fossil collecting was not a way to riches.

The winter of 1926–27 passed. Jim worked on his father's farm and I went to visit my cousin in New Mexico. By spring Jim and I were still determined fossil hunters, but more for the pure pleasure of finding the remains of extinct forms of life.

After our education about the improbability of getting rich collecting fossils, Jim and I were again blessed with greenhorn luck. Jim's cousin told him about some big bones weathering out in another small canyon on his father's ranch. The big bones turned out to be an articulated or semi-articulated skeleton of a long jawed proboscidean that was later described by Harold Cook as *Trilophodon phippsi*.

Jim and I spent all of the summer of 1926 working out this deposit from which we collected an articulated skull and jaw and the major parts of the skeleton of one individual and the skull and parts of the skeleton of another. In addition to the proboscidean material, we collected the remains of several different animals related to horse, camel, and carnivore.

We finally interested the director of the Colorado Museum of Natural History in Denver, Dr. J. C. Figgins. He sent his colleague Harold Cook to see what those kids had found in Nebraska.

In the meantime, Jim and I had heard how the fossils were first soaked with shellac and later encased in plaster of Paris. When Harold Cook arrived, he saw our collection. Some specimens were still in the canyons; the remainder was in storage in Jim's father's barn. Our collection, financed only by the spare cash I had accumulated working at various jobs, represented our months of work.

I happened upon the archive letters from Cook, Figgins, and Childs Frick and they make interesting reading. Harold Cook was impressed with our collection and knowledge.

How did Cook and Figgins arrange to get enough money from interested people to pay Jim and me for our summer's work? I don't know, but we were pleased. Jim and I ended up with $1000 each.

Jim went to high school and finished in three years and I kept my share in reserve and matriculated at the University of Nebraska in the fall of 1927.

I first enrolled in the college of Engineering and took whatever courses were necessary for freshman engineers during the first semester. Back then, all male students were required to take ROTC so I was required to march and follow the usual military drill. This was a big pain to me in view of my previous experience in the Civilian Military Training Corps (C.M.T.C.) at Fort Des Moines, Iowa.

After the first semester, I became better oriented and wanted to register for a course in astronomy. My advisor was an understanding instructor who told me that I could not take astronomy as an elective as long as I was in Engineering College, as there were too many mandatory courses to take. He suggested that I might take it when I became a senior. I objected and took a night course in astronomy anyhow. My advisor then told me that I could re-enroll in the Liberal Arts College and then I could take whatever I wished. He reminded me that I should stick to his plan to receive my B.Sc. I continued to take the courses that interested me. I had little thought about future employment. I just wanted an education in the academic world that seemed more important to me than one narrowed down to one profession only."

Marie explains, "I am sure that when Morris began his autobiography just three years before he died, he summed up the important things in his life and that was to learn all he could about what interested him. He didn't change when I think back on the things he did. He always said that when he learned something then he was satisfied and didn't care whether he passed it on by publication or not."

Morris' autobiography is how he saw himself. The remainder of this book documents how others saw Morris F. Skinner. Here are just a few of his many accomplishments.

- In 1973, in addition to being named Frick Curator Emeritus, the University of Nebraska named Morris a research affiliate of the Nebraska State Museum.

- In 1978, the University of Nebraska conferred upon Morris the honorary degree of Doctor of Science.

- In 1985, the Society of Vertebrate Paleontology presented Morris and Marie a certificate that reads, in part, "…for a long career of excellence in collecting, curating and studying vertebrate fossils including the documentation and interpretation of the stratigraphy of the sediments in which they occur. Together you rank among the great collectors in the history of vertebrate paleontology."

- In 1986, the University of Nebraska awarded Morris the title of "Alumnus of Distinction."
- Morris' bibliography includes 25 titles and at his death, he was working on two more. One had to do with his research on the horses of the world and another on the Oligocene geology of North Dakota.
- In April of 1989, and in the year that Morris died, he wrote, "For the past seven years I have studied the increasingly important ground water conditions in this area. Along with this, I have ongoing studies of the streams and drainages of the Niobrara River, from eastern Wyoming to its drainage into the Missouri River. These studies may never become the printed word but I make my findings available to anyone who is interested and keep notes and charts ready for exhibition and use."

As those who knew him would have expected, Morris F. Skinner, then well into his 82nd year, was still in relentless pursuit of knowledge about the complex character of the Earth.

Chapter 2

Shirley Marie White Skinner

Flypaper? After seventy years, Marie remembers that her first encounter with Morris was through a maze of flypaper.

In 1930, flypaper came in a small cardboard cylinder about the diameter of a 35mm film canister and possibly twice as long. When you removed a paper cap from the cylinder, you exposed a thumbtack attached to the end of a coil of paper rolled up inside. The paper, impregnated with a sticky substance, was irresistible to flies. You put this flycatcher into use by sticking the tack into the ceiling and pulling down on the cardboard cylinder. The spiral of the sticky paper curled out of the cylinder and when totally exposed, hung down like a long spit curl. Depending upon the height of the ceiling or other anchor point, and the height of the person who had to avoid them, these sticky fly-traps reached down to about shoulder high. In those days before air-conditioning, but in days when many Nebraska ranch house windows were without screens, one flypaper strip was never enough. The Frank Lessig ranch house was filled with such streamers on a hot August day in 1930.

Some twenty-one years earlier, on a stormy twenty-second of March in 1909, Shirley Marie White was born into a troubled family who lived on a small farm in the Highland Grove area, about eight miles north of

Ainsworth, Nebraska. She was the youngest, the last of three girls and a boy.

Her father, Lloyd B. White, had been born in Wisconsin in 1859. He was a railroad man who had helped construct the Portland-to-Portland railroad. The ribbon of rails on this Maine to Oregon run traversed Nebraska. Lloyd White was a foreman of a crew in charge of laying tracks and building railroad trestles through Nebraska, including the masterpiece over the Niobrara River near Valentine. This portion of track named the Fremont, Elkhorn and Missouri Valley Railroad, later became the Chicago and Northwestern. Now the rails are gone and the rail bed has been converted to a hiking and horseback riding trail.

Lloyd lived in Fremont, Nebraska but he liked what he had seen to the West in Brown County, near Ainsworth. He also liked what he had seen of Maria Schumann, a fully trained nurse who, in 1894, had found her way from the Russian Steppes to the little frontier town of Fremont. She was of Viennese and Lithuanian stock and was born in East Prussia. She departed her homeland because she had grown to hate European monarchies. Most of all, she hated the German military. On Berlin sidewalks, she often had to suffer the indignity of stepping aside for Prussian officers.

Although ten years separated their birthdates, Lloyd and Maria were married in 1899, in Fremont. In 1904, they brought their three children, William, Florence and baby Esther to the Highland Grove area in Brown County, north of Ainsworth.

Lloyd worked hard learning to be a farmer and was active in the country community. He promoted schools and his name is on the first record of trustees of the Highland Grove Methodist Church. He was a militant prohibitionist and voted a straight Republican ticket.

Maria's ambition like her strength was boundless. She raised a garden and tended to her chickens and milk cows. The butter and eggs she sold in Ainsworth paid for clothes and luxuries such as the *Youth Companion* magazine and the *Christian Herald*. If Maria had thought about politics, she would have labeled herself a liberal.

In 1909, little Shirley Marie was born to parents now fifty and forty years old. A crippling form of arthritis had attacked Lloyd and he was no longer up to the demands of farm life. His nephew, Lloyd LaFarge, who was weak of mind but strong of back, came to do the heavy work. Disappointments and differences were already taking their toll on the White marriage when little Shirley Marie was born.

Shirley Bayles, a very pretty teenager, came to help the infant girl's mother. At the same time, Lloyd's brother Fred, a bachelor, came to visit and see his new niece. Fred was smitten by Shirley Bayles and made no bones about it. One day he was flirting through an open kitchen window while she was doing dishes. To show her contempt of his courting, she threw the pan of dishwater at him through the window. Lloyd took such insolence as an affront to the family name and immediately named a newborn calf "Shirley," and forbade the use of that name for his baby daughter. From that early day, Shirley Marie White became known to the world as "Marie," a tribute to her mother. Marie's two sisters also went by their middle names, again by their father's choice. So, as in many homes that have a formal living room and another room where most living takes place, "Shirley" was relegated to formal occasions and "Marie" is the name she has lived in.

In another four years, the rolling plains of Brown County reminded Maria, more and more, of her bleak future in the Russian Steppes. Her marriage ended in divorce. Maria spared the children the details of what all led to the separation. She only told the children that she didn't like buying tobacco for the men out of her butter and egg money.

The children later moved back to Fremont, Nebraska with their mother. Maria returned to her nursing profession. She bought the Fremont Hospital, with assistance of a large mortgage, and ran it in conjunction with a training school for nurses. Education was important to the single-mother and Fremont offered her children access to good education. In spite of the hardships, she put her three daughters through college.

The first three years at the University of Nebraska were disappointing to young Marie. She admits that her scholastic efforts there were sometimes second to her active social life. She had also come to realize that she was not equipped to reach her first goal, to be a teacher of Latin.

She had translated all of Virgil's *Aeneid* with the help of a "pony". A "pony" is now often called a "study guide" such as *Cliff Notes*. It is not cheating to use a "study guide"; but most professors can spot their telltale traces through your work and will reward you with a "C."

So Marie used a pony to translate Virgil's *Aeneid* and she was struggling with Ovid's poetry as well. Although Virgil and Ovid, written in plain English, cause most readers to struggle, Marie saw her efforts with those two Roman poets' ancient Latin as "writing on the wall." She believed she would never master Latin well enough to teach it. She discovered that her real interests were in the zoology classes and other Earth Science courses that dealt with evolution, genetics, ornithology, and botany.

She had spent from June to mid August, of 1930, in summer school to make up a failed mark in Ancient History. Although it made her a believer in the adage, "Those who do not learn from history are doomed to repeat it," she needed a rest. There was solace in abundance at the Frank Lessig ranch near Ainsworth. It was near her birthplace in the Sandhills and near the family farm her mother owned.

The family farm and surrounding countryside were dear to Marie. She knew every country road and admired the cottonwood groves and shelterbelts once planted by settlers. Farm and ranch buildings were often nestled in the groves and rows of trees, called shelterbelts, as they provided protection from the brutal winter winds. The Timber Culture Act, or Tree Claim, passed by Congress in 1873, offered a one hundred and sixty-acre "Tree Claim" to anyone who would plant trees and homestead. The trees, many planted as necessary proof to the govern-ment that a settler intended to stay, had now grown tall. The "proved up" land then is a patchwork of prairie, fields and trees.

To the south and west are the Sandhills, grassy knolls of dune sand interspersed with lakes, untouched by the plow and sparsely settled. The ranchers who live within the Sandhills call them "God's Cattle Country." The dome of Nebraska sky fills each day with glorious sunrises, sunsets and cloud formations. The night sky, so far from city lights, sparkles with stars. The whole area has not changed much since Marie was young. It was a haven to her then and is to this day. So in August of 1930, Marie returned to the Frank Lessig place to soak up solace.

One late afternoon, a bone digger and his assistant knocked at the door of the Lessig farmhouse. They had stopped by to ask permission to prospect the canyons or maybe just to fill their water jugs. When the young leader of the twosome stepped inside, Marie greeted him through the shelterbelt of flypaper strips that hung from the ceiling. Marie remembers a stocky, healthy, tanned young man with direct, blue eyes—and she remembers the coils of flypaper between them.

This was the instant when Morris knew he had met his match. From Morris' vantage point, he saw a lovely, young lady who was as irresistible to him as the flypaper strips were to the flies. He may not have seen the flypaper strips at all, and if he did, it's doubtful that he would have remembered them seventy years hence. His direct, blue eyes met hers and stuck there. Morris said later that he knew immediately that Marie was to be his. From that day on, Morris clung to Marie like flies to fly-paper. He was hopelessly trapped although Marie, as of then, did not know it.

Morris cast aside his fossil hunting enthusiasm and spent the next two weeks intensely courting Marie. He made daily visits to the Lessig Ranch to take Marie out in the evening and to bring her back to the ranch by dawn. They never went on dates to local dances, but instead Morris took her to wherever he was camped and they talked the night away. It was obvious to Marie that Morris was serious about their relationship. Marie wasn't sure, in two short weeks, that Morris was the man she really wanted.

On September 14, Morris arrived at the Lessig Ranch in his Model-A Ford and escorted Marie back to Lincoln and the University of Nebraska. Marie introduced Morris to her family before they both pursued their fall courses. Morris arrived at Marie's house each day to take her to class and returned her home each night. They spent every waking minute together on the weekends.

Marie's family and friends were alarmed at Morris' everlasting presence and spoke plainly, and often, about it. They believed that Morris and Marie were not suited for each other. Their main concern was that Marie was only one year away from getting her degree. They could see that Morris was hard working and studious, but his rough edges were forever topics of their criticism. He was on the University of Nebraska wrestling team, but wrestling was rough-and-tumble behavior and only dubiously accepted as a gentleman's sport by potential in-laws. Marie's family didn't know that Morris came from good New England stock and was well versed in social graces. Morris' good-natured but bawdy witticisms masked the gentleman he truly was.

Although Marie was twenty-one, her mother didn't count birthdays. She wanted Marie to finish her fourth year of college and could see that Marie's current preoccupation was jeopardizing that goal. Marie's mother issued an edict that the two were not to see each other for a while.

Morris and Marie decided if their proximity around Marie's family caused such undying contempt, then they would take their togetherness somewhere else. They began to meet on campus between classes and visited the library together after school.

Utterly blinded by his affection for Marie, Morris could not see how difficult this arrangement was for her. Marie obviously had to choose between Morris and her family. Marie finally suggested that because of all the family objections, they should break up.

Such a notion stunned Morris. He didn't drop to a knee and romantically ask for Marie's hand in marriage, but he blurted out a proposal on the spot. "Let's get married!" was his obvious solution. When Marie didn't concur immediately, Morris added that if Marie wouldn't marry him, nothing would matter anymore. He would give it all up, quit school and just go home and be a cowboy.

Such an ultimatum stunned Marie! Morris, at that time, had become modestly famous. His unusual fossil hunting career had spawned an article and photo by a fellow journalism student in the *American Magazine*. The picture showed the intrepid young explorer, Morris, in a canyon setting, and the two-page text told of how he was paying his way through college by selling the amazing, prehistoric things he had found. All of his future, or so he said, was now riding on Marie's shoulders. She was also aware that Morris was not a good horseman and presumed that cowboying would not be an easy life for him. As Marie had already come to take this ever-present young man for granted, she allowed her feeling of being stunned to be a feeling of being needed. So, she said, "Yes."

Within a week, their blood tests were complete. As a blood test was the only key to wedlock, they could see no point in further waiting.

Bright and early on the morning of October 3, 1930, and only slightly over six weeks since she and Morris had met, Marie put on her best dress. If she had had more time to plan, her best dress might also have been a white dress and befitting of a beautiful bride. Her best dress, though, was a black dress. Nonetheless, it did have a large, white cape collar. She grabbed her schoolbooks and hurried off to campus although this wasn't her intended destination. On the way, her English teacher, Louisa Pound, offered her a ride to campus and they had a nice conversation about everything—everything except what was on Marie's mind.

She met Morris and they quietly drove to Marysville, Kansas, some seventy-five miles south of Lincoln.

One of Nebraska's biggest exports to neighboring states, in those days, was weddings. Ainsworth, Nebraska lovers, who wouldn't or couldn't wait, drove one mile across the border to Wewela, South Dakota if they were in a real hurry or another eighteen miles to Colome if it wasn't urgent. Morris and Marie had all the time in the world so they drove seventy-five miles from Lincoln, to Marysville, Kansas.

Judge Potter rounded up a witness while Morris dug into his pocket for a dime store ring, and the more precious three dollars to pay the judge. Marie's third finger, left hand, sensed danger, saw the ring coming, and swelled in defense. While Morris tried to urge the ring over her knuckle, the judge solemnly pronounced, "These are the ties that bind."

As Morris pushed, twisted, and finally forced the ring into place, he uttered, "Well! It should be binding if *this* is the way it goes."

Marie could not contain the nervous energy she had reserved for the solemn moment of their wedding vows. Morris' comment struck her funny. She began to laugh helplessly all the way to the car and most of the way back to Lincoln.

Back in Lincoln, they met as often as possible while keeping their marriage secret. Her family finally accepted the inevitability of their togetherness, although Marie couldn't find a way to break the news of their marriage. She moved out to work for her room and board in the wealthy part of town. Morris and Marie returned to Ainsworth for Christmas break and still kept their marriage a secret.

Although morning sickness is a biological function, Marie discovered it in her Botany class instead. On a cold January day, she found herself unable to look through a microscope at a green plant with a yellow bloom.

Marie was apprehensive about telling her mother and two sisters, but, to her surprise, she discovered that they were relieved. They had suspected it all along. She had needlessly suffered weeks of torment.

Now that the news of their marriage was out, Marie's next consuming worry became whether little Barbara, inside her, could hold on for dear life for a respectable nine months. Barbara did, and with a day or two to spare.

Meanwhile, Morris had rented the upper floor of a house with a Murphy bed, a kitchen and a bath. It was their first home together, and it was where they began their unforgettable job of learning to live together—and this they did for the next sixty years, with or without Murphy to referee.

Marie was Morris' wife, his assistant, his secretary, his housekeeper, and his friend. Morris and *wife* got along magnificently. Morris and Marie of the other labels often fought with fervent enthusiasm. Their arguments were loud and purposeful—never about family matters or marriage or over things like, "Do you really love me?" They argued about science, and the proper words to use in their papers, and how things should be said or done.

An ongoing argument was about listings. The standard, throughout the profession, was for fossil collectors to arrange their lists according to the taxonomic order of animals. This meant that lesser forms should be listed first and higher forms listed later on their field lists and in their publications. Marie insisted on conforming to these standards. However, Morris knew that Mr. Frick's emphasis was on camels and his own emphasis was on horses. For that reason, Morris insisted on placing the Frick Lab emphasis at the top of the lists, where they could be easily seen, and placing the lesser forms of animals, such as reptiles, as the bottom dwellers. Often these life long debates rose to full-fledged arguments. Their voices would resonate throughout the hallowed halls of the American Museum of Natural History and most always within shouting distance of fellow workers. Fortunately, their office was on a floor away from the galleries open to the public. A museum visitor then could only presume that the stereotypical, quiet, deliberate, study of serious, spectacled, scientists was taking place off in the wings.

Morris and Marie were not ones to promote stereotypes. When things got too rambunctious one of their colleagues might step in to tell them, "We can hear you out in the Big Room!" The Big Room was a large, storage and work area that was adjacent to Morris and Marie's Tower Room office. Their Tower Room was like a castle tower and it offered a commanding view from North through South of Central Park and Manhattan. Malcolm McKenna, their friend and colleague, pronounced their relationship "the longest divorce proceeding on record."

Malcolm just didn't understand where Morris and Marie drew the line. Morris and Marie, man and wife, got along splendidly. Morris was simply intolerant of his headstrong assistant and secretary. Marie was simply intolerant of her headstrong boss. But, as to their marriage? Their marriage was everlasting. Once that dime store wedding ring popped over Marie's knuckle, it bound them together like flies to flypaper.

Chapter 3

Childs Frick and the Frick Laboratory

Although this book is about the fieldwork and collecting that Morris so loved, this account of his life would not be complete without mentioning the Frick Laboratory where Morris and Marie spent every winter.

The life in the laboratory, in the center of Manhattan Island, was a far cry from the rustic life on the Great Plains. Nevertheless, if it were not for Childs Frick, who paid for the expeditions, there would be little to write.

Childs Frick and Morris Skinner came from very different backgrounds. In spite of their differences, the thirty-five year employee-employer relationship between these two men was one of enduring loyalty, trust and respect.

Childs Frick (1883–1965) was the son of Henry Clay Frick, a partner of Andrew Carnegie in the Pittsburgh steel and coke industry. He had a consuming interest in paleontology and geology and especially the study of the evolution of mammals. He contributed large funds to the collection of Cenozoic fossils throughout the Great Plains and in Florida, Alaska, New Mexico, Arizona, California, and Nevada as well as the countries of South America and China.[7] Many individuals and institutions also benefited from his philanthropic acts.

The employees of Childs Frick were certainly not in it for the pay. Mr. Frick financed many undertakings. He was frugal and expected his employees to include, as part of their paycheck, the intangible benefit of being part of a unique scientific endeavor. This austere employment package worked for some. The starting wage for a field assistant was only a few dollars per day. It was six dollars per day for Tom Lucas in 1942. It was six dollars per day for me in 1960. In spite of long days and harsh working conditions, some of us eagerly went back to work each collecting season. We endured the meager pay for the greater opportunity to head for the badlands again.

The members of the Frick Laboratory were divided into two groups, those who received salaries from the American Museum of Natural History (AMNH) and those who received salaries from the Frick Corporation in Pittsburgh. Employees who received a Pittsburgh paycheck were: Morris and Marie, Ted Galusha, Loren Toohey, Otto Geist and all the field men: Joe Rak, Jack Wilson, Will Chamberlain, N. Z. Ward and field assistants under Morris and Ted.

Both groups worked under Mr. Frick's direction, but there were inequities. AMNH people received holidays and pay raises whereas Frick Corporation people received no vacations. Mr. Frick assumed that his employees were on vacation in the field. Unless an AMNH person told them of a raise they had received, the Pittsburgh paid employees never got one. That meant asking Mr. Frick for a raise and he would usually *think* about it. He was always displeased at the question. Morris managed to get raises, but Marie worked for about ten years at $225 per month before an increase. Marie said she never had the nerve to ask for a raise; she thought, as others did, that she was lucky to have a job at all.

---

[7] Ted Galusha wrote a comprehensive report on the Frick collections in *Curator*, March 1975, Vol. 18, No. 1.

Then when summers came and field collecting began, Marie spent many weekends typing without any pay at all.

Although her associates will disagree, Marie felt her usefulness in the field was minimal. She claims that she was no good at finding fossils and never collected one unless it was small and durable. Except for some of the first summers when they were doing quarry work, Morris even did most of the camp cooking. In spite of her perceived lack of fossil collecting skills, she loved going to the field. Her joining us on an expedition often meant that someone had to ride in the back of the pickup. It was a small price to pay to have Marie come with us.

When Morris asked Mr. Frick about retirement benefits, Mr. Frick replied, "You'll have to take care of your own pension." Mr. Frick did not approve of Social Security either and so they were the last to pay into it.

That is why the Skinners bought Nebraska farmland. Morris was thrifty by nature so the financial conditions of working for Mr. Frick were taken in stride. Morris and Marie managed to save enough to make down payments on farms out of a salary that kept most people living from month to month. They bought good land cheaply in the 30s and when they had paid for it, Morris studied the bond market and went into that.

I asked Marie if it would have made a difference if Mr. Frick had paid them more for their services. She replied, "In some insignificant ways, perhaps. It is hard, though, to imagine living a life much different than we did. Morris and I did nearly all of the things we ever wanted to do and in ways that we wanted to do them. Searching out the secrets of the natural world and sharing our lives with the caliber of men and women who contributed to these pages was our biggest reward of all."

PART 2

MEMORIES OF THE MEN

Photo 6: Sioux Co., NE (1947) Tom Lucas, Morris Skinner, Leonard Nelson, Bill Laverty, Fritz Skinner

Throughout the half-century of Morris Skinner's fieldwork, a host of men went to the field with him. His first associate, in 1926, was Jim Quinn. During the 1930s, Morris hired Ralph Mefferd, the Potter

Brothers, Gordon Fletcher and Howard Williamson. During those years of depression and drought, they collected fossils in canyons and tributaries of the Niobrara and Snake Rivers of Nebraska. In the late 30s, they moved westward to Gordon, Nebraska and on into Sioux, County. This migration was generally upstream, along the Niobrara River drainage toward its headwaters in eastern Wyoming. In the winters of 1937–38, Howard Scott Gentry and Albert Potter joined Morris in Arizona to collect the fauna in Papago Springs Cave.

In the pre-World War II years of the early 40s, the earlier associates were phasing into other occupations. Morris and Marie's young son, Fritz, then began his fifteen years of fossil collecting. Tom Lucas worked the summer of 1942 and the two summers of 1946–47 after he returned from military service. During the summer of 1943, Morris hired John Beattie and Bill Lear. Leonard Nelson and Bill Laverty joined the crew in the late '40s.

Loren Toohey's fieldwork with Morris was during the early '50s. Carl Elfgrin and my brother, Bob Emry, came on the scene late in that decade. By then, the emphasis was the fossil localities of northwestern Nebraska, eastern and central Wyoming, Texas and the Dakotas.

Part 2—Memories of the Men is about these men and others who went to the field with Morris. My time with Morris began in 1960. My recollections will follow in *Part 3—Bone Digging Days.*

Chapter 4

James Harrison Quinn

(1906-1977)

Jim was one of seven children of Clayton and Mareka Thompson Quinn of Ainsworth, Nebraska. He and Morris Skinner began their love of paleontology together. Jim found "giant bones" in the canyons of his parents' farm and the two young men's curiosity led to the careers that became their lives. Their friendship endured for over half a century. The two septuagenarians were together, hunting fossils, when Jim died.

The *Society of Vertebrate Paleontology News Bulletin*, Number 112, 1978, published two articles about Jim Quinn. Jim authored the first one, on pages 45–47, and entitled it *Nebraska Revisited*. The second, on pages 53–54, Morris Skinner wrote about his lifelong friend. It is entitled *James Harrison Quinn 1906–1977* and, tragically, it is his obituary.

The two articles are included here. There is one exception. Some time later, Morris penned a final paragraph at the conclusion of Jim's obituary, in his copy of the *SVP News Bulletin*. It concludes this chapter.

Nebraska Revisited
By James H. Quinn

In 1926, Morris Skinner and I began collecting rhino bones from an outcrop in a canyon on our farm. Skinner said these bones were worth money. We sent a large jaw (we didn't know what it belonged to) to the Museum at Lincoln, Nebraska, to test its salability. Dr. Barbour sent us a check for $10.00 and said he hoped additional material could be recovered. We had refrained from mentioning that we had a hillside full available.

We got busy and collected four or five skulls, jaws and assorted skeletal material, boxed the bones up and waited for fortune to descend upon us. The value of rhino bones seemed to diminish with abundance for we were sent a check for $125.00 for the whole lot. We were very disappointed at this. Morris turned to seek his fortune elsewhere and I began contemplating a winter of corn picking.

My father's cousin was staying with us at the time. He had carefully watched the fossil collecting episode. One day he said to me that he had gone "bone hunting" and had found a very large bone. He thought of "cashing in" on the bone hunting activity, but since they had turned out to have little value, he would show me the bone and I could do as I wished with it.

I dug a hole in the sand alongside the large bone and encountered a skull and jaws of the long jawed or four-tusker early Pliocene mastodon, which we called *Trilophodon*. Skinner on hearing the news decided we ought to investigate further on the grounds that the mastodon might be more salable than rhino bones, said to be very abundant anyhow. We uncovered the skull, jaw and leg bone, photographed them and since it was late in the fall covered them up again to wait for spring.

We had conflicting reports concerning the value of our discovery but by spring were ready to do some collecting. We assumed the whole skeleton would be there, in the sand, so we arranged to uncover an area of sufficient scope for the purpose at hand. With two shovels, we worked from dawn till dark for four or five days digging down to the level of the bones. These were situated part way up a very steep slope extending from the canyon floor upward to the foot of a vertical cliff. The area excavated was roughly triangular—about 30 feet on a side. There was nothing on the "floor" of the "quarry" except the skull and leg bone (femur) we had seen initially. We finished the clearing of the area at about sundown on a disappointing day. The very last shovelful of sand, which I removed just to make the deepest corner of the excavation neat, exposed the tip of a mastodon tusk.

The next morning we began again—having learned nothing—to uncover another large area by the side of the one finished. After about a week of shoveling sand down the hillside, we had worked down to about two feet above where we thought the bones ought to be. Then the cliff behind us decided to fall onto our excavation so we had to begin again to clear that material away. Finally, we had the area cleaned to within two feet of the bones and went to work on the job of exhuming a skeleton still unseen and unknown. Just back of the tusk was a lower jaw and on the end, the skull lying on its side and all articulated. Behind the skull, the nearly entire skeleton was piled as though by a whirlpool. A humerus stood almost on and wedged between the jaws. We tried hard to get it out without breaking anything but couldn't seem to manage it. Then we tried to break the humerus in two and that didn't work either. The two of us pushed and pulled until finally it came loose. There was no breakage. The success of our collecting with both the rhino and the mastodon material was not a matter of skill but the indestructibility of the bones themselves.

Fifty years later, I visited the two sites. I could hardly recognize either. The back walls of both quarries had weathered down, filling in the quarry floors so that very little physical indication of our labors remained. Some sizeable trees had grown in both areas. The talus or debris slopes formed by the "spoil" we had shoveled down slope below the quarries had washed away.

The mastodon stands in the Denver Museum with one of the rhinos. There are one or two rhinos in the Nebraska museum together with a musical instrument (the bonafone) made of rhino ribs by Henry Reider. (These are all constructed from the $125.00 worth of bones Nebraska got from us in 1926). I set up one skeleton in the Field Museum in Chicago. I do not believe the American Museum exhibited any of their share of the rhino material, although they have a large quantity of it. The mastodon in Denver is still the most complete single individual insofar as I know and happily will long outlast the hole in the hillside from which it came.

Incidentally we may have done a bit better on the Mastodon, market wise, because the Director of the Denver Museum (Colorado Museum) at that time found $2,000.00 for us in exchange for our summer's collection. We were given to understand they were not "buying" the collection but merely remunerating us for our summer's work. I don't suppose we cared much what they called it. We could use the money.[8]

---

[8] *The Society of Vertebrate Paleontology News Bulletin,* Number 112, 1978, pg. 45-47

Obituary
James Harrison Quinn 1906–1977

By Morris F. Skinner

James Harrison Quinn was born in Ainsworth, Nebraska, November 27, 1906. His career in paleontology began in the summer of 1926 when he and Morris Skinner pooled their interest in rocks and "bones" and went fossil hunting in the canyons of Plum Creek on Jim's farm near Ainsworth. Their first find was a large jaw that they immediately identified as a giant fossil "lizard" just like the dinosaurs that Barnum Brown found in Wyoming. At that point in their career, time and space meant nothing and they were just as thrilled to learn from E. H. Barbour at the University of Nebraska that the jaw was from a rhinoceros. Barbour wrote, "If you find more of the animal it will be much more valuable." "Valuable" was a word that meant money as much as science to two very poor 19-year old boys. Beginner's luck was with them when they dug in again at the site of their first find for they uncovered a fantastically rich deposit of well-fossilized bones of a short-limbed, heavy-boned rhinoceros that required no special care in removing from the matrix. In fact, when struck, the ribs rang like crystal and were used by Henry Reider, a preparator at the University of Nebraska, for his "bonaphone," an instrument Henry modeled from the xylophone.

The first summer's work yielded enough cranial and postcranial elements for the Nebraska State Museum to mount a composite skeleton (NSM 1169) that Barbour referred to *Teleoceras fossiger*. One of their more complete finds was a semi-articulated skull and mandible of

Trilophodon phippsi now mounted in the Denver Museum of Natural History.

At the end of the summer, Jim and Morris decided that they both needed more education if they were to pursue their new career. Morris enrolled at the University in Lincoln and Jim, at age 19, enrolled at the Ainsworth High School. Three years later Jim graduated valedictorian of the class of 1930 with a straight 96 percent average. During summers from 1927 through 1929, Jim and Morris prospected the Niobrara River canyons and contiguous drainages near Ainsworth with impressive results. They had frequent visits from Paul McGrew, Ned Colbert, A. L. Lugn, Harold Cook and Barnum Brown, all of whom had a decided influence on Jim's future.

In 1939, Jim started to work at the Field Museum of Natural History in Chicago as a preparator and assistant in paleontology. His enthusiasm and ingenuity found many outlets, one being a method for making plaster casts of fossils. In 1940, he published a classic paper on the use of liquid rubber as a casting medium for preparing brain casts. This was a great advance over the then known techniques of plaster piece-molds, flexible glue, or gelatin molds.

The death of Jim's wife, Margurite in 1947, left him with two young children, a son, James, Jr., and a stepdaughter, Dixie Lee. In the face of this tragedy, Jim responded boldly with the best of disciplined thought and action, and with gloomy predictions from almost everyone, took his children to Tucson where he entered the University of Arizona. Three years later, he received his B.Sc. degree. In 1950, Jim married Doris Glover and with his new family moved to Austin, Texas, where he entered the University. Jim's major professor in paleontology was John Wilson, and it was under him that Jim published in 1954 his thesis on the Miocene horses of the Gulf Coastal Plain. Jim spent the next 19 years at the University of Arkansas—the last eleven to retirement as Chairman of the Department of Geology.

Jim's primary interest was always fossil horses. In Arkansas, however, he turned his attention to fossil invertebrates and local geology. He published numerous papers on the occurrence, morphology and stratigraphic position of invertebrates and the geomorphology and physiography of the Ozarkan area. His bibliography includes 55 titles.

Jim's last paper in Fieldiana Geology, dedicated to Rainer Zangerl (June 24, 1977) was titled "Sedimentary processes in *Rayonnoceras* burial."

On a perfect late summer day such as bone diggers dream about (September 14), Jim and Morris decided to prospect the Snake River in Cherry County, Nebraska, and most particularly the type locality of *Alligator mefferdi,* in which Jim was interested. At the site, Jim spotted a partial mastodon tooth on a small ledge about 15 feet up on the outcrop and crawled out to retrieve it. As he crawled back, a rock from overhead fell without warning and knocked him off the ledge along with other debris. He died a half hour later of massive internal bleeding[9]

—

Jim's remains were cremated and I scattered them on the east side of Quinn Canyon with a granite marker where we both had such fond memories of our youth together. In 1955, Jim published the first comprehensive studies of Equid Lower Dentitions. In this, I cooperated *off* the record, as the Frick collection was not available at that time. I maintain his objective observations have stood the test of time along with his Phylogeny Chart (see *Miocene Equidae of the Texas Coastal Plain.*) 1954[10]

[9] *The Society of Vertebrate Paleontology News Bulletin,* Number 112, 1978, pg. 53-54

[10] The final paragraph is typed as Morris Skinner had penned it at the end of Jim Quinn's published obituary.

Chapter 5

Photo 7: From Sheep Mountain, SD—July 1964

Ralph Mefferd

Ralph Leroy Mefferd was born on September 9, 1909 in Ainsworth, Nebraska, the son of Dr. Ira W. and Edna Goodrich Mefferd. His father was an optometrist, a watch and clock repairman, and a charter member of the local Lions Club. His three sisters (Edith, Georgia and Mildred) died of scarlet fever when Ralph was a boy and his mother also died when he was only fourteen. Ralph spoke of these tragedies in a way that made his friends think that he had difficulty coming to terms with his loss.

Ralph was tall and gangly with long arms and legs. He was called "Spider" by his school friends because he could make his limbs look tangled when he sat down. A nickname "Meff" stuck with him throughout his adult life. He had a natural dignity, decent manners and unassailable good humor.

He did well in school and graduated from Ainsworth High School in 1928 but did not pursue a college education. His second son, Charles, understood that he had visited a relative in California, where he had studied to be an electrician. Ralph's father married again and fathered three more girls. Ralph took any job available. One was as a ranch hand on the Davison Ranch in south Brown County, where he met his future wife, Dorothy Idora Dodds.

Morris Skinner, three years older than Ralph, hired him as his assistant on June 3, 1931, a summer that was spent moving overburden by long hours of shoveling and prospecting the steep Niobrara River and Plum Creek canyons. The high point of that summer was when Ralph and Dorothy were married, in Colome, South Dakota on August 31.

Ralph and Dorothy were products of the Great Depression. They not only survived the national calamity, but they thrived with only meager assets. They had two sons, Ralph Jr. and Charles. They raised their family in a small, modest home on the south side of Ainsworth and just around the corner from Morris and Marie. As times got better, they continued their simple lifestyle. To Ralph and Dorothy the basic necessities were not only sufficient but they were preferred. Ralph's father had taught him hunting and fishing. Consequently, he was a life-long sportsman and conservationist and was president of the Sandhills Rod and Gun Club.

Ralph was quick to learn the basics of fossil collecting—to recognize the skeletal parts and to learn the rudiments of stratigraphy. Morris was not given to easy compliments. Instead, he was apt to coerce perfection by outright faultfinding. Morris' critical tutelage often produced quick learning apprentices. Ralph was no exception.

Morris and Ralph continued their prospecting and fossil collecting throughout the canyons of northern Nebraska. Nearly all of these sites were within the tributaries of the Niobrara River.

Most of the tableland above the canyons was under the plow. That was long before irrigation canals or deep wells brought water to the crops. The farmers were utterly dependent upon the benevolence of Mother Nature. The "Dirty Thirties" were years of drought that have not since been rivaled. The drought was so widespread that the devastated country was given a name, "The Dustbowl." The sky was often dark but not because of impending rain. The overcast was caused by wind-born topsoil from far-off fields. The absence of rain was on everyone's mind and in everyone's conversations.

Somewhere below the rim of a Niobrara canyon and below the dryland fields, the drought was the subject of a friendly wager between two fossil hunters. As the red dust of far off Oklahoma drifted over on a stiff southern wind, two parched throats finalized their bets. That evening Morris Skinner recorded their pledge in the back of his 1935 account book:

> "Mefferd bets Skinner 10 bottles of good beer that during the year of 1935 there will be twice as much moisture as in 1934 at Ainsworth, Nebr. by the post office records.
>   Signed Morris Skinner
>   Ralph Mefferd."

There is no evidence in later account books to show who won. It may be evidence of a bet that was never claimed. I expect that the Post Office in Ainsworth can still settle this bet if heirs of either Morris or Meff wish to drink up their inheritance.

The 1935 account book reveals another controversy of the day. It also had to do with water. This entry states:

> "Skinner =
> Weight of fish   =   1 lb.
> Weight of tank and water   =   100 lb.
> Total weight   101 lb.
> Potter = The Above is not true"

This entry has no evidence that Meff weighed in on this difference of opinion. It appears that Potter had little confidence in his position. First, he does not make known what the truth should be and second, he does not wager 10 bottles of good beer.

During the summer of 1941, Ralph Mefferd found evidence that, once upon a time in Nebraska's past, there was no shortage of water. He found a well-preserved skull and jaws, with some postcranial bones, of an alligator. Meff found this specimen in sandy channel deposits at about the middle of the Ash Hollow Formation on the George Sawyer Ranch, in Cherry County, Nebraska. Charles C. Mook studied the alligator and published his findings in the March 11, 1946 edition of the *American Museum Novitates*.[11] He named this new species *Alligator mefferdi* in honor of the discoverer. This distinct honor is a source of pride for Ralph's sons who have copies of that scientific publication as keepsakes.

Ralph Jr. and Charles are not only proud of their father's scientific endeavors, they are also proud of his rugged, adventuring spirit. They

---

[11] Mook, C.C. 1946. A new Pliocene alligator from Nebraska. *American Museum Novitates*, no. 1311.

recall a time when their father and Morris were working on the Snake River in Cherry County, Nebraska. The Snake River was fittingly named. Not only did it wind out of the Sandhills in a serpentine style, but also serpents of the genus *Crotalus* were prevalent among its rocky outcrops. This creature was not called by its scientific name in those parts. When a prospector happened upon a *Crotalus*, its common name was often announced in tense timbre at the apex of an adrenaline-fueled jump. "Rattler" is the word that better describes *Crotalus*.

Ralph and Morris were leaving the fossil quarry where they had been working and noticed a rattler coiled quite close to the path. The buzz of a rattlesnake has a way of honing your senses. What may look like an ordinary cow-pie takes on startling new characteristics once your snake detectors have been aroused. When Morris and Meff began to look more closely, they didn't find just one rattlesnake. When they finished counting, they had estimated some seventy rattlers within a very small area!

The Mefferd boys remember the story. Was it true? Thirty years later Morris retold this story to a young, bone-digging apprentice who ventured with him back to the Snake River. Remnants of stovepipe and plaster were strong circumstantial evidence of veracity.

As we pawed through the bits and pieces, Morris told me that it was in the fall of the year and the numerous rattlers were gathering at their den in advance of winter hibernation. Once Morris and Ralph located the snake den, they waited for several days until they presumed that the migration had ended. They then plastered up the cracks that led into the den. They plastered a length of stovepipe into the crack that they believed was the main entrance and placed a barrel underneath. When warm weather returned the next spring, the awakening rattlesnakes crawled out of the stovepipe and dropped into the barrel. Morris recalled leaning his head over the barrel one day just as a rattler slithered out of the stovepipe. It narrowly missed his cheek. They put a damper in the stovepipe after that so they could turn off the flow of rattlesnakes when they were checking the barrel.

Once they captured the rattlesnakes, they crated them and sent them to the American Museum of Natural History. The crate of snakes, securely locked and posted with a warning "Do Not Open!", was delivered to the fifth floor of the 77th street building. The fifth floor was not a storage area nor was it a floor of natural history displays. The fifth floor did not fit your first impression of the word "museum." The fifth floor housed the offices of curators, curator assistants, secretaries and other living, and therefore snake-squeamish, people.

It seems that a janitor, for a yet unexplained reason, pried open the crate in spite of the lock and warnings! There is no scientific scale, such as the Richter, to measure the magnitude of this event. Nonetheless, aftershocks were immediately felt as far away as Ainsworth, Nebraska. Morris received a terse telegram from Dr. Noble, Curator of Reptiles, "HOW MANY SNAKES DID YOU SEND?"

Marie tells me that Morris wasn't certain of the exact snake count and informed Dr. Noble of the substantial element of doubt. I suppose that could be the truth—just could be. Knowing Morris Skinner though, he kept excellent records. What better way to give the New York City museum staff a taste of the West than for them to be forever cautious that genus *Crotalus* lurks nearby?

Marie said she could only imagine how the reptiles could hide around those old hot water radiators. She informs me that the box of *Crotalus* went on to scientific fame. They were the first rattlesnakes to be bred in captivity.

The working conditions in the field were hard to comprehend by some of the New York City museum workers. Morris often found dramatic ways to demonstrate the contrast. In addition to Morris and Meff's rattlesnake adventure, Morris told me of another time when east met west.

The museum laboratory where technicians prepared the fossils was on the seventh floor. One day Morris noticed some preparators who appeared to have nothing to do. He inquired about their idleness and they informed him that the freight elevator was broken. The freight elevator was the usual route that fossils took from the basement to the laboratory. The basement storage had boxes and boxes of yet unopened fossils from the field. The preparators took them from their field condition, encased in plaster and rock, and turned them into museum quality specimens.

Morris recalled the many times that he had hauled large fossil casts out of badland canyons on his back. He saw the broken freight elevator as an opportune time to teach these city boys a lesson. He went to the basement and selected a specimen - not because of its rarity but because of its size. He strapped a cast of some two-hundred pounds to his back and muscled his way up seven flights of stairs. Upon delivering the cast to the wide-eyed preparators, he nonchalantly said, "There, that'll keep you busy for a while."

As seasons turned into years, Morris could see that Ralph was committed to their scientific work. Because of his professionalism, Ralph was chosen to take charge of work in the South Dakota Badlands when Morris and others worked elsewhere. Ralph worked either alone or with an assistant at these times. One such assistant was Leonard Nelson, a hardened veteran of World War II, who said it was all he could do just to keep up with Meff whether prospecting or digging.

Another anecdote that Ralph's sons remember their dad telling them about was his fall from the side of Sheep Mountain in the South Dakota badlands. Charlie relates, "Dad had been working alone on Sheep Mountain. He had set his pack above himself on a ledge. As he reached for it, he lost his footing and slid down the side of the mountain almost a thousand feet. It almost shredded his clothes…and lots of bumps and bruises, but no broken bones."

The uninitiated may think that the terrain features of the Big Badlands of South Dakota, such as Sheep Mountain, rise from the plains like the Rockies. Not so. The badland features are remnants of the plains themselves. The White and Cheyenne Rivers and their tributaries carved downward through the hundreds of feet of a vast plain of volcanic dust and fluvial sediments to expose the fossils that were buried within. The mountains are simply those mesas and spires that have not yet been washed away. Many of them are joined by durable, blade-like ridges called clastic dikes. Taken in total, the badlands look strikingly similar to the surface of the moon. These features aren't visible from far distances as typical mountain ranges are. When you arrive at the edge of the drainage system, there stretched out before you in startling beauty are the mountains of the Big Badlands.

Sheep Mountain rises about six-hundred feet above the wide valley floor. You may wonder then how Meff could encounter a precipice of "almost a thousand feet" from which to fall. It can be explained to a level of scientific probability in at least three ways:

- The Theory of Elasticity: Meff could have fallen six-hundred feet and bounced four-hundred more.

- The Theory of Relativity: One can easily conclude that Meff's fall from a badland spire could have been one thousand feet, relatively speaking. The height of Sheep Mountain is relative to the following three conditions: (1) The temperature. (2) The amount of drinking water you carry. (3) The distance you must walk out after you have taken a tumble. If on that day, the badland temperature was the normal one hundred in the shade, if Meff carried no water, if he had to haul his abraded carcass along for miles after his one-man avalanche, then Sheep Mountain was Everest, relatively speaking. Meff was a typical bone digger so he most surely encountered all three conditions.

- The Theory of the Expanding Universe: Mountains I have climbed in the past are much more difficult to climb today. A six-hundred foot mesa in Meff's day is undoubtedly a one-thousand foot mountain by now.

My irreverent humor may camouflage the seriousness of Meff's fall. Working alone in the badlands is a serious business and even a modest disabling condition could prove fatal. Meff's fall wasn't just some "war story" that he passed on to his sons. Morris told Loren Toohey about Meff's fall many years ago. To quote Loren, "Meff was working an area where a Protoceras channel was exposed and because of the rubble, he lost his footing on a steep slope. How he came to stop his bouncing and sliding, I don't recall. It might have been his geology hammer or a concretion that saved him from going over the edge and hence to his death."

During the winter months when the fossil hunters returned to the museum in New York City, Ralph's gift for precise illustrations led him into illustrating camel and carnivore specimens at Mr. Frick's Millstone Laboratory on Long Island. Morris took great pride in a detailed map and illustrations that Ralph drafted for his report on the Pleistocene fauna of Papago Cave in Arizona.

Loren Toohey has vivid memories of Meff's outstanding penmanship. Loren writes, "I think Mefferd had a mechanical lettering device implanted somewhere between his head and fingers. In 1938, Mr. Frick wanted more camels from the famous Hemingford camel quarry and Schultz, University of Nebraska State Museum (UNSM), agreed to Mr. Frick's request that allowed Morris to work the sites in October and November. I was sort of the UNSM representative because Schultz trusted very few. I learned many tricks of the trade from Mefferd, Fletcher, Potter and Morris. Anyway, UNSM numbers were assigned to specimens and duly recorded in the standard UNSM record book. The first time I watched Mefferd enter the data I was surprised and amazed at the uniformity of each number and letter as well as the spacing. The old straight pens needed uniform pressure to avoid wide and narrow lines or circles. He was great! On the outside of the Skinner garage in Ainsworth is one-half of a packing box lid. There is Mefferd's printing as precise as ever. Morris must have brought it back from New York as a reminder of past days."

During World War II Ralph left the Frick Laboratory and took employment with the Civil Service for several years (1943–1946). The Ainsworth airport was being revamped and enlarged. It became an Army Air Corp air base where fighter and bomber crews were trained. It provided jobs for many of the local men including this writer's father. If the old hangar is still standing, you can look high into the ceiling trusses and see "Ken Emry, 1942" scribbled there in carpenters' chalk.

For a short time after the war, Ralph again worked in the fossil localities with Morris. This employment was indeed short. On October 4, 1946, Morris wrote in his diary, "Before I left for camp I may have gotten Mefferd lined up with the new irrigation project for the Ainsworth area as a draftsman with Clyde Burdick."

Burdick was the local area chief of the Bureau of Reclamation. Ralph's familiarity with the landforms being studied was so needed that he was soon hired as a geologist for the United States Bureau of Reclamation. His knowledge of sedimentation and stratigraphy lead to his nomination for membership in the Geological Society of America.

Ralph Mefferd's bright career was all too short. Meff was only 55 when he died of a heart attack on Christmas Day in 1964.

Nearly a quarter century later, when Morris Skinner was in the last month of his life, he said to Marie, "If I had kept my old crew I could have gone anywhere and done anything—Mefferd, Fletcher, Potter, Williamson."

Chapter 6

The Potter Brothers

Jess, Albert and Hugh Potter all worked for Morris Skinner in the mid-1930s when Burge Quarry on the Snake River in Nebraska was in full operation. The Potter brothers were as different as the skills required in a fossil quarry.

Quarry work requires a range of expertise. You go about this work as many people eat layer cake, by quickly consuming the cake and leaving the icing layers for last. Quarry workers eat the "cake" or overburden away with picks and shovels and sometimes dynamite and machine powered scrapers to get to the frosting, the layers that contain fossils. This preliminary work does not require excessive training. The best trait for removing overburden is a muscular build.

You treat the icing with much more care. The fossils themselves are like nuts in the frosting, to be carefully removed and savored. You must be able to make tentative identifications of the exposed fossilized bone before you proceed. A curve of exposed bone, to the untrained eye, may appear to be a four-inch piece of rhino rib that holds little scientific value. The untrained hand is tempted to thrust a geology pick behind it to gouge it from the matrix. If temptation commands a swing of the pick, the sinking feeling of pick shattering hidden fossil is a lesson not soon forgotten.

The trained eye will see the curve of bone as an exposed section of a lower jaw. The trained eye has near x-ray abilities to visualize the entire specimen. It completes the picture beyond the few visible inches and sees the whole jaw and the articulated skull as well. This x-ray vision allows the trained bone digger to judge the extent to remove the overburden to bring the specimen into full view.

The trained eye of a bone digger is sometimes fooled. Once you remove the overburden and you carefully whisk the sand away with delicate instruments, a four-inch piece of rhino rib with little scientific value may be all that remains. Nevertheless, there are the times when your whiskbroom exposes the envisioned jawbone. Then, you lean down and blow the sand away to see that first glimpse of upper molar. You know the skull is there! It was there since long before man walked the earth and you are the privileged one to see it for the first time. When you have reached that level of training, bone digging is in your blood. Paleontology may never be your career, but you will most likely be a bone digger for life.

Quarry work requires a range of expertise and temperament and the Potter boys filled all roles. The Potters, as did all other bone diggers, started with the cake before they got that first taste of icing.

The older brother, Jess, was a steady worker, quiet and serious. He filled his place on the crew with little fanfare, but he was dependable and valuable to the effort.

The younger brother, Hugh, was Jess's counterweight. He was fun loving and talkative. He shoveled overburden without complaint and became adept at the finer aspects of quarry work too. When Morris went prospecting, he always enjoyed having Hugh along. Long after Hugh's bone digging days, he would stop by the Skinner's home just to talk. Marie states of Hugh, "If I say he was cute and sweet, those words just come automatically. He is still a good friend and when I happen to meet him in Valentine or we talk over the phone, my day is the better for it."

The personality of the middle brother, Albert, was somewhere in between. He became a life-long paleontologist, so the bulk of this story naturally belongs to him.

Albert was born on May 12, 1913 in an upstairs bedroom of a ranch home near Valentine, Nebraska. He attended a one-room, rural grade school and graduated from Valentine High School in 1930.

Albert had some training in paleontology before Morris hired him. Albert's first serious encounter with a fossil was at the age of sixteen when he and his brother were riding fence on their ranch. They discovered a bone that the University of Nebraska identified as belonging to a new mastodon.

The discovery led to a continual string of museum crews to the Potter ranch. Whenever paleontologists visited the Sandhills, the boys tagged along and they were often hired for their ability to use a shovel.

When Paul McGrew, a paleontologist from the University of California Museum, stayed at the ranch, he hired Al as his field assistant at the Gordon Creek Quarry near their house. It was just a matter of time before Al and his brothers pursued their interest in fossils to the Burge Quarry where the Frick Laboratory began working under the direction of Morris Skinner in 1934.

Fossil quarries are not only filled with sand and fossils; they are filled with the conversations of workers as they go about their tasks. Whenever anyone asked Al where he was born, he replied, "I was born upstairs." He was referring to the upstairs bedroom of their ranch house. Nonetheless, Al's demeanor often led one to believe he was noting a difference between his status and that of the questioner. Those around him could see that Al was interested in expanding his circle of influential friends. Al's "upstairs" birth was his first stepping-stone on his way to success.

After several seasons under Morris Skinner, including the winter of 1937–38 in Arizona, Al entered the geology program at the University of Nebraska. As a student there, he served as an assistant preparator in

Morrill Hall for the Nebraska State Museum. Following his sophomore year, Al transferred to Chadron State Teachers College in western Nebraska and was also an assistant at the fledgling museum there. He graduated in the spring of 1941 with a Bachelor of Arts degree in biology and continued his work at the college museum.

During the next few years, he joined his earlier mentor, Paul McGrew, on expeditions to Honduras, Guatemala, and Panama.

In 1946, Albert joined the faculty at the University of Northern Iowa and taught science education. He received his master's degree in education in 1951 and continued postgraduate work at Duke University and the University of Colorado. He continued to teach at the Malcolm Price Laboratory School in the Department of Teaching at the University of Northern Iowa until his retirement as Professor Emeritus in 1978.

Following his retirement, he continued his love of fossil hunting and visited many of the old quarries. His two manuscripts on Miocene beavers lay unfinished at his death. Albert Potter died on December 7, 1997 of prostate cancer.

His family name lives on. Childs Frick honored the Potter brothers, and especially Albert, for their help in those early years in Burge Quarry. A new species *Pseudoceras potteri* firmly anchors the Potter name in scientific circles.[12]

---

[12]Albert Potter's obituary was published in the *Society of Vertebrate Paleontology News Bulletin*; No. 173; pg. 74-77. Many of the facts in this essay are directly from his obituary but paraphrased.

Chapter 7

Gordon Fletcher

Gordon Fletcher was born in 1913 in Burke, South Dakota. He was the last of ten children born to a blacksmith and a mother of English heritage. Few wealthy families immigrated to that neck of the woods to increase their fortunes. Therefore, the immigrant parents of Gordon's mother were most likely poor, decent and hardworking.

If Burke, South Dakota was typical of most farming communities in the northern plains, the Great Depression came early and stayed late. Poverty was an accepted way of life. Gordon's mother, with nine other children to care for, could not have been ecstatic over Gordon's arrival. She may have even extended similar sentiments to the arrival of the preceding nine. According to Gordon's wife, Thelma Barton Fletcher, her mother-in-law was very cold in her manner and not receptive to any kind of merriment.

In due time the Fletchers moved some eighty miles west-north-west to White River, South Dakota where the Bartons were then living. The Bartons were the antithesis of the Fletchers; Thelma's parents saw each child as precious. Bringing their children to adulthood was a blessing. Mrs. Barton (nee Jessie Fink of Creighton, Nebraska)[13] was a very

---

[13] *Pioneer Stories of Brown, Keya Paha and Rock Counties, in Nebraska,* Published 1988 by the Brown County Historical Society, Edited by Shirley M. Skinner, Printed by Star-Journal, INC, Ainsworth, Nebr. 69210, Pg. 558.

beautiful, dark-eyed woman, prematurely white-haired, hospitable and completely family-oriented. This is where Gordon found his place of happiness, his acceptance into a loving family.

Thelma Barton Fletcher's grandparents had homesteaded on the Niobrara River near Meadville in 1883. Meadville is at the river crossing about twelve miles north of Ainsworth. Reuben and Esther Barton's homestead "consisted of 400 acres, mostly timber and grazing land, as his main business was raising horses and cattle. He brought purebred mares and stallions planning to raise purebred stock. All the mares but two were stolen so he slept at the picket-fence pen to keep thieves from stealing his valuable stallion. He also had a herd of cattle, but when the drought of 1893–95 was over, he had traded most of his herd for feed. He would take his team and hay frame, tie a steer behind and lead it to the hay flats about ten miles north, and trade it for a load of hay."[14] Such is the plight of the family farmer. Nonetheless, "Flour was 50 cents for a 50-pound sack and coffee was ten cents per pound. With wild game they managed to have plenty of food."[15]

Thelma's father, Lewis, "was born to Reuben and Esther Barton at Meadville...on the farm where he grew to manhood and attended school. On March 25, 1906, he married Jessie Fink. They resided in Keya Paha County for six years and then homesteaded and lived for 21 years at White River, South Dakota."[16]

Thelma was born at Meadville, in 1915; her early memories are of South Dakota. South Dakota was where she met Gordon Fletcher. Thelma's family moved back to Meadville, Nebraska in 1934, and the year that she graduated from high school.

Although the Bartons were a prominent pioneer family on the Niobrara River, the drought of the 1890s had set them back.

---

[14] Ibid. Pg. 557.
[15] Ibid. Pg. 557.
[16] Ibid. Pg. 558.

Unfortunately, the drought of the 1930s and the Great Depression were yet to come. The consequence was decades of hardship and subsistence living. Thelma's family moved to White River out of necessity and their move back to Meadville in 1934 was likely for the same reason. "They operated the tourist park and lived in the old log house that had been a part of the Mead Inn until it burned down May 7, 1936. About 5:30 P.M., Jessie Barton started the gas range to prepare supper. She left the house to shut up the chickens and upon starting back, she saw the whole west end of the building was a mass of flames. Neighbors gathered and firemen from Ainsworth and Springview were summoned, but arrived too late to save any part of the building or its contents. All the Bartons had left was the clothes on their backs."[17]

Thelma's father died in 1941 after a short time in Wyoming. Her mother and brother Don went to California in about 1942. Don Barton served in World War II and returned to a career in real estate. Don once remarked about his business, "If I hadn't made money in real estate then, it would have been because I was stupid." Stupid he was not. Don donated to the city of Ainsworth, Nebraska the funds necessary to build the Sellers Barton Museum (log cabin).

There is nothing left of the Meadville that was once located south of the river and above the bridge. There wasn't much to Meadville, Nebraska even then. The tourist park of the '30s was nothing more than a hamburger stand, a dance hall and a few primitive log cabins. A garage and gas station rounded out this little community. If you ran the dance hall, you lived there. If you ran the garage, you lived in the back.

1937 was a year of significance. In 1937, the Skinners bought their farm and lifelong home on the south side of Ainsworth. 1937 also marked the year when Morris hired Gordon Fletcher. In November of that year, Gordon and Thelma were married.

---

[17] Ibid. Pg. 558.

Thelma was an asset in the bone digging camps at Burge and Xmas Quarries. She furnished the fun and laughter, loved all the children and was a support and a good friend to the other wives.

For every married fossil hunter there is a spouse who is, at a minimum, tolerant of the profession. Camp life in the '30s was no picnic. Luxuries were nonexistent as the drought and Great Depression affected fossil hunting too. Thelma was born into a life of hardship so conveniences were never had and therefore never missed. She was self-confident, a dependable friend, and a hard worker. Although this story is intended to pay recognition to Gordon Fletcher and his time with the Frick Laboratory, Thelma was very much a part of Gordon's life.

Gordon was medium in height, slender and well built. He had sandy hair, blue-eyes, and was pleasant looking, but in an unremarkable way. You had the impression that he was all beige-colored. He was always pleasant in his manner, but that, too, was unremarkable, perhaps because he was so passive. He almost never expressed his likes or dislikes about any topic. He did not argue much, even when friendly arguing was a constant in camp life. In that regard, Gordon Fletcher was different from Potter, Mefferd or Williamson. The other three had to have their say about everything. Gordon stated his opinion on only a few topics. When he spoke, the others paid attention. One such topic was the proper setup for scraping overburden from the fossil quarries. Gordon was the expert in mechanics so he seldom accepted advice from his less informed critics.

Gordon's one distinguishing feature was an up tilted nose that Morris immediately thought deserved the nickname of "Flipperbill." Therefore, Gordon became "Flipperbill," "Flipper" or "Flip." Morris was not unfair about nicknames as he gave himself the Lakota Sioux name of "Hosay Tonka" meaning "He of the Big Butt."

One day, in Ainsworth, when it was raining too hard to work, the men got plaster and water and went to the hog-shed that Morris had converted into a place to pack fossils. With plaster, water, and time on their hands, they had the ingredients for making a plaster cast of something. Why not someone's head? By now, it should be obvious that it was Gordon, of course, who was the voluntary victim.

They greased Gordon's head and hair with Vaseline so the plaster would not stick and proceeded to wrap it with plaster-soaked burlap. They hoped that the plaster mold would be complete and without the slightest detail missing; so they placed soda-straws into his nostrils. These rudimentary snorkels allowed him to breathe as the plaster set. When plaster hardens, it gives off heat. The heat is not excessive but it is likely apparent and perhaps downright uncomfortable if your head is encased in it. As the plaster hardened on Gordon's head, the experimenters became bored. What sort of reaction might they cause if they pinched the soda-straws shut? So, they laughingly and quite sadistically found out. Gordon's contortions efficiently spelled out "P-A-N-I-C"— his clue in their game of charade. When they all agreed that they had guessed Gordon's pantomime correctly, the perpetrators opened the breathing tubes again. When the plaster was hardened, it was carefully cut away and used as a mold for the final cast of Gordon's head.

When I worked for Morris in the '60s, I came face to face with a plaster head that was under a workbench in the packing shed. It stared out with blue eyes that had been colored in with carpenter's chalk. Everything else was age-yellowed plaster—all beige colored. As I pulled it from its hiding place and mused over it as Hamlet once did poor Yorick, Morris said, "I see you found ol' Fletcher" and told me the story.

Marie believes that the cast went to California where Gordon later lived. She attests that when the cast was fresh, it was a perfect likeness of Gordon Fletcher. When I worked for Morris, Gordon had been haunting the packing-shed for about thirty years. Time had removed most of what could be called a "flipperbill."

Gordon was a valuable assistant, a hard worker and a friend of his fellow workers. He served the Frick Lab well for his eight years of employment. He was careful and responsible and was good with his hands, whether it was with carpentry, fossil preparation or machines.

It was Gordon who helped Morris install two transmissions in a 1928 Model-A Ford pickup. They needed something to pull a scraper to remove overburden from their quarries. They also needed a vehicle that could haul their specimens across miles of sandhills to Ainsworth. It was impossible to get a machine down into the canyons of the Snake River where the Burge Quarry was located, so they rigged up a series of pulleys and cables from a scraper in the quarry to a flat area of land above the canyons. When they hitched the Model-A pickup to the cable they had to race the engine to keep it from stalling. That only served to spin the tires in the loose sand. They outfitted the pickup with oversized balloon truck tires and that kept the wheels from digging in. Unfortunately, the pickup's little four-cylinder engine didn't produce enough power to turn the balloon tires when the pickup was hitched to the cable. Another modification was needed. They acquired another transmission that was identical to the existing one, cut out a section of driveshaft and mounted it inline with the first. Now the pickup had more combinations of gears than they could use. Each modification seemed to earn the pickup a new nickname. She was sometimes known as "Lulu Belle," sometimes the "Iron Mule," sometimes "Whoopee," and sometime "Maudie." "Maudie" or the affectionate "Old Maudie" became her preferred moniker.

Maudie became the tractor they needed. The driver could put both transmissions into double low and, even at idle, Maudie's drive wheels would turn in slow motion, the cables would tighten and, below in the canyon, a scraper would easily move overburden from the quarry to a dump pile. During the years that Gordon was employed, he always drove Maudie for the scraping work. Morris said that Gordon had a sixth sense for shifting gears to pull the scraper back or to go forward to scrape and unload.

One day the crew was traveling the unpaved road from Valentine, Nebraska on their way to the Snake River and Burge Quarry. They came upon a road crew who had ridges of gravel blocking the roadway. An arbitrary time on the workers' watches had signaled coffee break. Traffic was light on the road. It was so remote that in reality traffic was nonexistent, except for the bone diggers' two vehicles. Once the road crew had set their mind to having a break, the arbitrary time became as rigid as stone. The museum crew waited for a while and then grew impatient. Gordon was in the lead driving Maudie. Morris and Meff were following in their pickup. A polite "Ahooogah" on Maudie's horn only resulted in one of the crew waving and taunting, "You guys in a hurry?"

With that Morris commanded, "Give 'er hell, Fletch!"

Gordon selected a combination of gears that if labeled would read "Give 'er Hell!" He tweaked the spark advance lever to keep the engine from stalling and floored Old Maudie. Gordon and Maudie went spinning off through their piles of gravel as if they were not even there. Morris and Meff followed behind on the path carved out by Maudie. Morris said that the looks of surprise on the faces of the crew were something he never forgot.

Morris once told of a worried rancher in the area who had seen truck tire tracks going up and down sand dunes where trucks were not capable of going. The rancher had worries of a ring of cattle thieves with secret technology. He slept better after he happened upon the Frick Lab crew and Maudie.

During my years with Morris some thirty years later, Old Maudie was retired. She had her own special shed at the Skinner residence in Ainsworth. Her balloon tires only remained round because they had been filled with sawdust. Her pickup box was loaded with rolls of cable and other bone-digging paraphernalia. Every so often, we would pull her out of her shed and put a battery and a gallon or two of gas in her.

After some coaxing, she would sputter, wheeze and finally cackle to life. We would take her on short spins around town.

Morris used a piece of pipe as an extension for the short gearshift lever of the rear transmission. If it were removed, the modification that gave Old Maudie her secret power was not readily noticeable. Morris removed the pipe from time to time to befuddle the unwary. A friend of the Skinners was visiting and was reminiscing about driving a Model-A in his youth. He wasn't aware that Old Maudie was not an ordinary Model-A. When Morris finally coaxed Maudie to run, he secretly selected reverse on the rear transmission and removed the pipe. He asked the fellow if he wanted to take her for a spin. Of course he did. His every possible selection, except reverse, of the front transmission resulted in Old Maudie lurching off backward whenever he released the clutch. Morris roared with laughter and chided the fellow for forgetting all he knew about driving a Model-A.

Healthy outdoor living does not make you immune to the ills of the human race. Even rugged Morris complained about fleeting aches and pains and was forever trying to find remedies. Gordon developed swollen, painful hemorrhoids while working near Meadville and Thelma thought they might respond to Iodine. Iodine was a favorite medical aid along with Mercurochrome back then. Iodine is not often found in medicine cabinets today. For those who have had it applied to a cut or abrasion it is a very warm memory. It was said by those in the know that Gordon ran all the way to the Niobrara River!

Gordon and Thelma went to New York City in 1938 and 1939 where Gordon prepared the fossils that the crew collected in the summertime. Their first child, Ronnie, was only 3 months old when they drove into Manhattan to look for a furnished apartment. Of course, they needed a cheap place to live. Gordon and Thelma drove to 84th Street where Thelma waited in the car while Gordon walked up and down the block looking for vacancy signs. By some strange chain of circumstances, Morris and Marie drove right up behind them. Imagine the relief and utter joy that two "river rats" had when they bumped into their old camping buddies on the streets of New York City!

Gordon left Frick employment on June 15, 1942. He was greatly missed. He went with the Bartons to California where Gordon had wartime employment at the Northrop Aircraft Factory. He then found lucrative work at the Metro Goldwyn Mayer movie lot as a prop man.

Although Gordon and Thelma drifted apart, he remained a life long friend of his brother-in-law, Don Barton. Don recalls, "Gordon was a prince of a guy…When I came home from World War II, it was Gordon who seemed to understand my feelings and apprehensions better than my own brothers…He seemed to know what we had gone through and that it took time to settle down to a peaceful civilian life…I admired the guy." Gordon Fletcher died in June of 1998.

Chapter 8

Photo 8: Snake River Falls, Nebraska—July 1959

Kids in Camp

During the seemingly endless decade of the thirties, the fossil collecting camps were also the summer homes for the fossil hunters' wives and children. During the Great Depression, luxury had disappeared from people's lives. The years of drought made the hardiest souls long for a day of blue sky, free from windblown dust. Yet, those days of camp life spawned treasured memories. They are the first memories of Morris and Marie's daughter, Barbara, and son, Fritz. Barbara recollects that her youngest years were not boring days of hardship at all, but days of fun, camaraderie and mischief.

Barbara was born on August 17, 1931, in Ainsworth, Nebraska. When she was only a few months old, her father began to build a tiny cabin near Meadville, along the Niobrara River north of Ainsworth, Nebraska. He began construction during his Christmas break from the University of Nebraska. The 1931 Christmas break also marked the time when Morris discovered the renowned Xmas Quarry upriver, near Sparks, Nebraska.

Morris, Marie and Baby Barbara moved into their little home in the summer of 1932. It had few amenities beyond a roof. On May 6, 1933, Morris F. Skinner Jr. was born. Baby Fritz also spent his first years in the tiny cabin that was still a work in progress. Morris ended his diary entry that day with, "I got some boards in the afternoon and put mopboards and sills in the cabin." Marie recalls those years at Meadville as the best time of all for her family. They lived each day by making the best of what might come their way. Although the world was in crisis, "fun" is a punctuation mark that is scattered throughout their memories. Marie and Morris had little to live on, but Marie had two babies and the natural beauty of the familiar land near where she too was born.

Barbara's first memories are of watching older children at the rural schoolhouse nearby. She remembers the simple things, such as going to the little Meadville store and having the best of all treats, a grape soda pop, straight from a cooler of ice water—or sitting on the sandy beach of a hidden backwater, known as the Meadville Bayou, and watching people wade and swim.

When the Skinners visited Ainsworth, they stayed in a small house next door to Barbara's grandparents. In addition to suffering the hard times of economic depression and drought, Barbara came down with scarlet fever. She and her mother were quarantined together because of the contagious and serious nature of the disease. Barbara remembers being very sick and feeling the relief of damp, cold cloths that her mother put on her. She also remembers her grandfather and brother Fritz, attempting to entertain her and cheer her up through a closed window. When the infectious days had ended, Barbara recuperated on a pallet in the yard and watched her father and Jay Williams build the trailer house that would be their next camp quarters. A trailer house in those days was a novelty.

In about 1934, Morris began to collect fossils in Burge Quarry on the Snake River, some fifty miles west. The Snake River lies almost entirely within Cherry County, Nebraska and is miles from a major highway. Evidence of the Snake River's remoteness is the division line between Mountain and Central Time. The seemingly haphazard, dotted line on the map, jogs left and right to stay as far away from people as possible. It crosses the Snake River at Burge Quarry. The Snake is a fast moving stream of crystal clear water that flows into the Niobrara. Its cascade over the cap-rock of the Ash Hollow Formation is the largest waterfall in Nebraska and is fittingly called the Snake River Falls. It may have the distinction of being the least visited waterfall of its size in America.

Morris pulled his rustic trailer house to the Snake River behind a Model-A Ford pickup and Marie cooked for the crew on a Coleman camp stove. She even made pies in the little camp stove oven. The Fletchers, the Mefferds, and the Skinners were one big happy bone digging family at Burge Quarry. In the evening, Morris hooked up his radio to Maudie's battery and they would listen to the news. There were no local radio stations in those days, so the nighttime, free from solar interference, was the best time for radio reception.

Although those hot dusty years made fond memories, Marie is still astonished that Morris would ever think of taking them there. Burge Quarry, hours from medical help, was synonymous with "rattlesnake" for they were everywhere. This was near the rattlesnake den that produced the famous snakes that Morris and Ralph Mefferd shipped to the museum. Nearly everyone at the Burge Quarry encampment had a close encounter with old rattle-tail or "Sintehla" (siñ-TAY-khlah) in the Lakota Sioux tongue. When Morris went into the night to "water the lilies," one struck at him. It was convincing proof that rattlesnakes are nocturnal. They crawled under ground tarpaulins and lurked where they least expected it. The parents watched their kids closely and Marie remembers her constant yelling, "Come back here!"

Barbara, who was almost four, and Fritz, who had just turned two, slept in a tent right next to the trailer house. One night a ferocious thunderstorm came up. Barbara vaguely remembers her stark-naked father and Ralph Mefferd trying to hold their tent down. She remembers waking the next morning with her father sleeping beside them on their camp cot and feeling very safe. When the men came to the tent, they teased Morris, calling him "Mother Skinner."

The children went with the men to get water at the nearby ranch-house windmill. It was a treat because Barbara could play with Florency, a girl about her age. Morris also had bought a cow, "Jingle Faucets," and kept it at the ranch. He milked it each day so that the children would always have fresh milk. The bone digging party made do with three, ten-gallon cream-cans of water per daily trip. Their water ration depended on the wind blowing sufficiently to turn the ranch-house windmill wheel. The water was not only for drinking and casting fossils, but also for cooking and for washing hands and faces when they absolutely needed it.

On warm summer days, they would trek down to the river to bathe and play. Marie carried Fritz on her back. Thelma Fletcher watched over Barbara and Dorothy Mefferd took her toddler Ralphie. Barbara missed stepping on a rattler by only a foot on the path to the river. On another outing, a rattler swam beside them in the river with its rattles held high and dry.

On some evenings, the whole crew went to the Snake River in Maudie and played in the water while the men caught fish. When they cleaned their catch, the children gathered around to learn all about their innards. The men stripped the float bladders from the fish and gave them to the kids to play with. They were like tiny balloons that they could pop when they were dry. The children were amazed at how many eggs a female fish could carry. They fried the fish eggs along with the fish and they were always a treat. I too remember eating fish eggs from the female fish that we caught in the river and bayous near our home. We enjoyed them too and I suppose we should have. Although poverty was our lot in life, we were rich with a world-renowned delicacy called caviar!

Barbara remembers the rainbows after summer showers that she, Fritz, and young Ralphie Mefferd hoped to follow to the pot of gold. Ralphie rode a stick-horse all day long, making it buck and gallop. They played in the sand of the quarry, building anything that came to mind. Who doesn't remember the soothing effect of cool damp sand on buried legs and feet? Barbara states, "The best plaything we had was our imaginations and there was no limit to what we could come up with."

Although Morris had responsibility for the whole camp, he made sure he spent time with his kids. When Barbara was four, her father tried to teach her to count to five. Virginia Cummings, a playmate, was five years old and she could count to ten. Barbara laughs that she must have been a miserable failure as her mother tells her she would count on matchsticks, "one, two, five, free" and her determined father would throw up his hands and exclaim, "Jesus Christ!" He would tutor her with even more resolve. Nevertheless, being frightened by her father's tenacity and outbursts, she never ventured much beyond, "one, two, five, free" even if Virginia Cummings could count all the way to ten.

Marie told of the many times that Barbara created mischief by sharing the scuttlebutt she had heard in one corner of the camp with the butt of the observation. When Thelma Fletcher quietly told Marie that she didn't much care for Howard Williamson's girlfriend, Barbara made sure Howard knew what Thelma thought. When Morris went cottontail hunting, he bagged more than the Skinners could use. Barbara watched him dress the cottontails. When Morris asked her to take some to the Fletchers, she trotted them right over. When she gave them to Thelma she innocently added, "Father saved the good ones and gave you the ones he cut the bad parts out of."

Barbara remembers growing out of her shoes in May and going barefoot until school started in September. Marie acknowledges that the children often went barefoot but doubts they were ever shoeless all summer long. One summer Morris even bought each of his children a pair of Sioux Indian moccasins. Barbara still has hers, holes and all. Nonetheless, Barbara remembers her summers free from shoes.

A diversion for the Skinner kids was to sit and watch the men work. Their cousin Howard Williamson, with his pipe going full blast, would signal Gordon Fletcher who drove Maudie back and forth pulling overburden off the fossil quarry. They waited in great anticipation for the dynamite blasts that Morris ignited in the quarry to loosen the more stubborn rocks.

One day their grandparents drove all the way from Ainsworth to visit. Marie bathed Barbara and Fritz in a dishpan and dressed them for the occasion. The kids never ever felt deprived or fearful for their safety in camp. Barbara exclaims, "It was just great the way we lived. We had the attention of all of the adults and always had a lot of fun with them."

The search for fossils moved to Gordon, Nebraska for the summer of 1937. They rented a house in town and Morris prospected in that area. Gordon Fletcher continued to work for Morris. Gordon's wife Thelma's advanced pregnancy was a constant worry to little Fritz. He kept telling his mother to tell Thelma to have a bowel movement because her stomach was so big.

During the winter of 1937–38, the Skinners went to Arizona instead of to New York City. Morris prospected there and collected the Pleistocene fauna from Papago Cave, about five miles southeast of Sonoita. The children's grandmother lived with them then. Barbara was in the second or third grade and, being conditioned to harsh New York or Nebraska winters, insisted on wearing a winter coat to school. She was too proud to get angry with herself when the afternoon temperatures soared and she suffered; she got upset at the temperature instead. She enjoyed shooting marbles with the neighbor children and discovered the fun in winning at games. For Christmas, Barbara's grandmother made a dress, underwear, shoes and socks for a doll that her parents had given her. Best of all there was also a little doll bed complete with bedding and quilt that her grandmother had made. Barbara still has the doll-dress and quilt.

The Skinner's city accommodations during the Depression and World War II were not a vast improvement over a bone digging camp. They lived in a first floor apartment with the bathroom some two flights up. In New York during the War, apartments were scarce, especially for just a winter's stay. Their iceman, Frankie, a man with swarthy European features, carried large blocks of ice in his metal tongs to their apartment. He learned the news of the neighborhood as he made his deliveries. Therefore, Frankie would keep them informed of apartment vacancies. As Barbara and Fritz were young and often without a caretaker at home, they spent many days at the museum with their parents. Some friendly museum guards helped look after them.

Barbara remembers December 7, 1941 very well. On this day of infamy when the Japanese attacked Pearl Harbor, her father and brother were at sea. They had gone twenty-five miles into the Atlantic Ocean, deep-sea fishing. Barbara remembers the fear she felt until they were safely back home. Now that America was at war, they put blackout curtains on their windows and practiced air-raid drills at home as well as in school.

That Christmas, three weeks later, Morris managed to buy a Christmas tree and a goose at a market. Barbara's grandma cooked the goose and saved the fat to put on toast with salt and pepper. Barbara knows it was an exquisite treat as she has thought of this delicacy for half a century.

As Morris and Marie's office was in a tower at the corner of the museum, they had a commanding view of Central Park. They allowed Barbara and Fritz to play in the park as long as they remained within view of their office windows. Their parents warned them to watch out for strange people in the park. If anyone looked at them in a funny way, they were to scream and run away as fast as they could. They did this a few times although a stranger's look might not have qualified as "a funny way."

They loved Central Park, enjoyed playing hopscotch and hide-and-seek and developing their fighting skills. Barbara's descriptions of the battles with her brother are convincing. If their combat was the magnitude of Barbara's memories, they both could have won a black belt in the martial art of knock-down-drag-out. Fritz's teeth marks were reminders Barbara carried for years.

Barbara remembers going to the movies and watching Sonya Henne skate gracefully around the ice. When she received her first ice-skates, Barbara took them to the rink in Central Park and pretended to do everything that she supposed Sonya Henne could do.

The grace of her skating apparently fell short of Sonya Henne's example, as her parents quickly enrolled her in a ballet class. She and her friend Jill would ride the subway to Greenwich Village and take their ballet lessons in The Little Theater. After the lesson, they would go to a Chinese café and have tea and almond cookies. They would be so stiff and sore from the stretching of the ballet lesson that getting back home was a trial. They wouldn't practice ballet all week so they would face the same misery of sore muscles every Saturday afternoon.

Although Barbara claims that she and Fritz preferred to fight, they also had sweet times. They would shovel snow from neighborhood stoops to earn a nickel or a dime. Once they took their money to a flower vendor and bought their mother a nice bunch of daffodils for Mother's Day. Sometimes they would have a penny left over to put on the trolley tracks. If not, they flattened bobby pins under the wheels of the trolley.

When Barbara and Fritz dug up a handful of dirt in Central Park and smelled it, they agreed that Central Park dirt smelled old and used up. They longed for their home in Ainsworth where the dirt smelled fresh and clean. When the winter was over, they welcomed their return to the pristine beauty of their beloved Nebraska.

During the summers of the early years of World War II, the Skinners camped in a line shack on a ranch near Scottsbluff, Nebraska. Barbara was eleven or twelve and Fritz was about nine. Fritz went to work with his father most days and Barbara helped her mother. Barbara remembers those days near Scottsbluff as a time when she and Fritz perfected their fighting skills. They had few diversions or toys, so Morris braided them some whips out of old boot leather. This had the same effect as giving a mad bomber a supply of dynamite. They put their quirts to good use chasing each other all over the Sioux County prairies. Barbara writes, "Fritz and I did have fights that lasted for days. We became experts at hitting each other in just the right spot to make one of us cuss or howl with pain."

Although my words might depict the Skinner children as uncivilized, I presume that many of you who have siblings will agree that Barbara and Fritz were not alone in this eternal struggle. Fighting may be a more common memory between siblings than brotherly love.

Whenever the Skinners went into Gering, Nebraska for supplies, Barbara and Fritz would spend nearly the whole day at the swimming pool. They then joined their parents to pick out a comic book that had to last them for a whole week. Sometimes they had read every page before getting back to camp. They often ate at a Chinese restaurant, sometimes went to a movie, but always visited an ice-cream parlor where they enjoyed the best malted milk ever.

Morris caught kangaroo rats for their pets and Barbara and Fritz made them nests in oatmeal boxes. The kangaroo rats loved to stuff their cheeks with dry oatmeal. Barbara and Fritz would put their pets under their shirts where they would crawl around, tickle their bellies, and sometimes deposit the oatmeal from their stuffed cheek pouches onto their stomachs.

The following summer they lived in slightly better accommodations near a windmill. Leonard Nelson now worked for Morris and he was the source of Barbara's first girlhood crush. Fritz taunted and teased her about being in love with Leonard. To prove Fritz wrong, Barbara ended his torment by throwing a cup of coffee on Leonard, thus ending all hope of the one-sided romance gaining a second dimension.

Marie did the laundry in a washtub. She used a funnel-shaped device on a broomstick handle, to agitate the clothes. When she had rinsed and wrung the water from the clothes, Barbara would climb up the windmill and hang the laundry on the braces of the windmill tower. Fritz sometimes joined Barbara on the windmill tower where they would wait for cattle to come to water. They then tossed rocks at the bulls in hopes they could make them mad enough to fight. Sometimes they succeeded.

That was the last summer that Barbara went to camp. She stayed at the Ainsworth farm with her grandmother and helped raise chickens and do other chores. Although she missed her parents and brother, she loved her grandmother very much. Barbara affirms that she would give anything to return to those wonderful times when she was a kid in camp. Her feeling of safety, the fun of making do with very little, the joy of stopping on a hill or bluff for a prospector's lunch of warm cheese, tashupa sausage on plain bread, and a drink of water from a burlap-wrapped jug.

Chapter 9

Howard Williamson

Howard Houston Williamson was born on October 16, 1911 in Basset, Nebraska. Fate has an interesting way of choosing birthplaces. Tumbleweeds bump along on the wind and sometimes bunch up in a fence corner where a new colony of tumbleweeds takes root. Children are sometimes born to parents who traveled far and met in similar fortuitous circumstances. When Howard Williamson's father was a young man, he was such a traveler. Luther Lee Williamson was a Texan. He tumbled north on a cattle drive from the Texas panhandle and came to rest in a fence corner of northern Nebraska.

Luther's grandfather and father also wandered far from their homelands. They moved from North Carolina to Texas in the 1860s. Luther's father, Fredrick, owned a large cattle ranch near Hereford, Texas. Luther and his six brothers all received their degrees in "cowology" on their father's ranch. In the early 1900s, Luther settled down behind a bunch of northbound cattle and arrived in Springview, Nebraska. At the end of the long trail north, he met and married Ezoa (Zoe) Phelps. He never went back to Texas. Luther, Zoe and their family moved to Ainsworth where Luther worked for his brother-in-law, Fred Skinner. Fred had married Zoe's sister Ezada.

Zoe and Luther had three children: Frederick, Howard, and Genevieve. Howard was the least robust of the three. Frederick, the ambitious strong willed older brother, retired as a vice-president of Associated Booking Company of Chicago, Illinois. Howard's younger sister was of similar demeanor. Genevieve Hall retired as a teacher in Castro Valley, California.[18]

This story is about the middle child, Howard. He was of medium height, always thin, swarthy skinned and had dark eyes and hair. His Uncle Fred and Aunt Ezada had a son, Morris Skinner, who was some five years older than Howard. In small towns, relatives often look out for each other. Howard looked to his older cousin as a protector from the time they were small. If Howard didn't have a job, Morris would find one for him. When Morris needed help in his fossil quarries, it was only natural that he would recruit Howard.

Howard's slight stature was no asset when it came to lifting and carrying the heavy casts and loads, yet he provided several essential services while working in the fossil quarries. The scraper in the canyon quarry and the Model-A Ford "tractor" above the canyon rim were linked together by a series of cables and pulleys. Unless the machines on each end of the cable were coordinated, they were at the least useless and at the most dangerous. Howard was good at ringing the gong to coordinate their movements. He could watch from a vantage point and signal the driver, usually Gordon Fletcher, to reverse course. Gordon would select a proper combination of gears on the dual-transmissions of the old pickup, affectionately called "Lulu Belle" or "Old Maudie," and the scraper in the quarry would efficiently remove overburden from above the fossil beds.

When Morris' wife Marie wasn't in camp, Morris appointed Howard to camp-cook duties. His menus and recipes, outlined in the next few pages, are evidence that he was an adequate camp chef. *Adequate* always

---

[18] *Pioneer Stories of Brown, Keya Paha and Rock counties, in Nebraska*; Brown County Historical Society; Edited by Shirley M. Skinner; Printed by Star-Journal, Ainsworth, Nebraska 69210. Pg. 696-697.

sufficed in a bone digging camp as criticism of the cook resulted in your immediate field-promotion to that camp chore. Seven little words, added to the end of an unsavory comment about the grub, could sometimes forestall a change in cooks. Unless you looked forward to cooking, you were quick to add the suffix "…but just the way I like it!" to any comment such as "This slumgullion is salty as Hell!"

Howard's talents favored brain over brawn. He was quick to see the humor in any situation and put it into verse or adage. He was truly the camp minstrel and philosopher. Working on the Snake River of Nebraska and later in Xmas Quarry were the two highlights of his fossil-hunting career.

He and Mildred McCormick were married on August 19, 1939, in Nebraska. It wasn't long before Howard had completed the circle that his father Luther had not finished. Howard and Mildred tumbled off southbound and came to rest only 50 miles from his father's boyhood home in Texas. He and Mildred spent the rest of their lives, just southwest across the Texas border, in Clovis, New Mexico.

He operated The Village Record Shop from 1942 to 1976 and emceed Howard's Bandstand, a late-night record and request show on KICA and later KCLV radio stations. He created Montgomery Mouse, Madeline and Rosalinda Skunk, imaginary characters on his radio show and a favorite among Clovis teenagers during the 1940s and 1950s.

Mr. Williamson became an agent for local musicians, booking them for dances throughout the area, notably The Blue Notes. He dedicated himself to helping children and listening to their problems through his radio show. He became a surrogate uncle, father, brother and friend to the teenagers of Clovis during those years on the air. Their parents loved him too. His brother, Fred, said that after visiting Clovis and talking to the people on the streets, he was proud to tell them that Howard was his brother. Of course, Morris and Marie were also proud of their cousin.

Howard retired in 1976 and operated an Indian jewelry repair shop. He had served as chairman of the Curry County Republican Party and on the state rules committee. He was a member of St. James Episcopal Church.

Howard was a heavy smoker all of his life. He contracted emphysema and died of it at the age of 75, in Clovis, New Mexico, on October 1, 1987.

The story doesn't end here as Howard's words live on. He visited his family and friends in Ainsworth, Nebraska during his senior years. Although he was mild in speech and manner, he could spin yarns about his days in the fossil quarries faster than most trained stenographers could put them on paper. Morris and Marie's daughter Barbara tried her best to catch the essence of his humorous tales. His philosophy of life shines brightly from the next few pages. Howard's anecdotes are as he told them then to Barbara Skinner Lamb.

Standard Operating Procedures of Camp Life

Selection of Chef

Pecking order determines this duty.

- Junior member is the Stooge.
- Stooge's primary duty is attending senior members.
- Primary function of the Stooge is to cook.
- Complaints registered by senior members relieve Stooge of cooking duties.
- The complainer becomes the new chef.
- Senior members all understand that they remain senior by not complaining to the cook.
- Quality of cuisine is therefore questionable but unquestioned.
- Howard Williamson remained the Senior Cook

Recipes and Methods

Breakfast: Authority—Hebrews 13-8; "...*the same yesterday, and today and forever.*"

Pancakes, eggs and coffee.

- For pancakes: Use mix.
- For eggs: Swish hot grease over top of the eggs with spatula.
- Don't turn eggs over or they squish.
- For coffee: If all members use sugar, it may be mixed in with the coffee prior to perking. If not, let them add their own damned sugar.

Dinner—Choose any or all of the entrees:

Slumgullion

- One can each of beef-stew—spaghetti sauce—pork & beans.
- Heat and eat.
- If you accidentally burn it, clean the skillet with sand, but not too well, as it will make the next batch stick.
- A large variety of canned goods and seasonings leads to endless possibilities for Slumgullion.
- Slumgullion cooked in syrup, although not totally delicious did not elicit complaints. Instead, it led to a greater appreciation of other variations of gullion.

Salmon Hash

- Open can of salmon into skillet.
- Chop it up with the edge of the can.
- Throw the can into the canyon as it attracts bugs.
- Blend in two eggs.
- Remove small pieces of shell.
- Crumble a handful of crackers into eggs and salmon and season.
- Fry until eggs are cooked
- Put onto plate
- Drown in catsup.

Salmon Gravy—tasty when served on leftover pancakes.

- Open can of salmon into skillet.
- Chop it up with the edge of the can.
- Blend some flour into about one cup of milk.
- Stir into salmon and cook until proper consistency.

Corned Beef on Bread—delightfully tasty dish (at least no one complained.) Don't give this recipe to local cattle ranchers as the Corned Beef came from Argentina.

- Follow directions for Salmon Gravy and serve over bread.

Canned Corn—an excellent side dish.

- You can cream it; you can fry it; you can serve it on bread with a little onion.

Interpersonal Relations and Privacy Issues

To thine own self be true—then thou canst be false to any man. Translated: "Don't say anything to anyone about anyone else—for they shall learn the truth and the truth shall make them flee."

- Look outside your tent before speaking about your neighbor.
- Say nothing good about one member's wife to another member's wife as it creates jealousy.
- Remedy: When second wife becomes resentful of first wife, say something nasty about first wife.
- Objective: Boredom in camp is detested. Camp life that is plagued with boredom makes work drudgery. When camp wives become loquacious about one another, it is the end of boredom.
- A bone-digging husband who is mad at another guy's wife—justifiably or not—is a more productive member of the group.
- Young children form an excellent liaison between their parents and other members. Their introductory phrase, "My Mama said..." will always gain an attentive audience.
- In order to complete this liaison, one should always ask the child, "What else did your Mama say?"
- For bathing (if bathing is deemed necessary), use natural resources if available...such as a quick dip in the Snake River. For

a more luxurious alternative, you can attach a platform of supporting timbers to four posts approximately eight feet above the ground, set a 55-gallon drum on the platform and attach a faucet and showerhead. Surround the four sides of the platform with a canvas privacy curtain. Prior to going to work in the morning, fill the drum with water. All 55 gallons will warm during the day in the sun.

- It is not wise to have more than two women in camp if this luxury is to be shared by all.

Mechanical Procedures and Equipment

Safety when using tools:

- Don't throw your hammer into the canyon if it's the only one in camp.
- Don't throw an axe down hill to someone else.
- Don't lay steel pry bars with their ends pointing down hill, on steep, pine needle covered slopes above the quarry.
- When delivering tools or equipment from the top of a canyon, throw upwards at a forty-five degree angle and yell "HEADACHE" real loud.
- Be prepared for the worst.

Uses for Bantam #2 Shovels:

- To dig fossils or instant latrines
- Grappling tool for climbing hills
- Brake for descending hills
- Stool, a dish, or a desk for writing.
- Headrest when siesta time hits.

- To beat out Morse code by hammering on its blade with a geology pick.
- Weapon to kill rattlesnakes and other varmints.

Precautionary Measures when using "Old Maudie"

- Do not sit in the middle of the seat as the short shifting lever for the rear transmission may inadvertently perform a "rectomotomy."
- When driving in reverse with the rear transmission in "high," don't exceed forty miles per hour.
- Associated precaution: While driving in reverse, do not attempt to transport poles longer than the bed of the pickup. Unexpected obstructions may cause the poles to disassemble the windshield and everything in between.
- Further precaution: Being attentive to public relations is recommended by the driver and passengers in Maudie, while driving cross-country in reverse. They should properly identify themselves in order that the locals may not question their own vision and sanity.
- One more precaution: When driving Maudie in reverse through underbrush and Maudie refuses further travel and this is accompanied by a more resonant tone to the exhaust, stop and check the muffler. If it is in a vertical position rather than the customary horizontal position, reverse course, retrieve muffler, wire it back on and continue expedition.
- When driving Maudie on the highway, the driver should tip his hat to construction crews before plowing onward through their manmade obstructions to vehicular traffic.
- Do not leave oilcans under rear balloon tires. Sharp oilcan stems and balloon tires are not compatible.

- Always carry a supply of copper wire (22 gauge, insulated and uninsulated). The insulated wire will replace wiring torn off by underbrush. The uninsulated wire will substitute for other Ford parts such as muffler brackets, etc. It also works as a bushing in the generator or starter when the bearings have disintegrated.

- Do not attempt to ford rivers such as the Niobrara unless adequate flotation devices are properly functioning and a rudder is firmly attached to the frame. If you do, you will need the following procedure.

- Motor and differential repair: Allow fifteen minutes to disassemble and twenty minutes to replace the differential. The additional five minutes are to find the bolts lost in the sand. To remove grease from mechanical parts, set fire to the grease with a blowtorch (Beware of prairie fires). When overhauling Maudie, turn her on her side. Mechanic A (Skinner) removes head bolts. Mechanic B removes pan bolts. Mechanics should coordinate work to determine whether the pistons are removed from the top or bottom of block. Lack of prior planning wastes time.

How Wade Quarry Kept Its Name

Morris Skinner named it Wade Quarry as bone diggers had to wade across the Snake River to get to it.

In October, late in the season, the weather was getting cold. The river water was like ice. Zeke Grubaugh, Gordon Fletcher and I were working it while Morris Skinner was on a South Dakota trip. We decided to hell with wading. We went back to camp, got a board, a .22 rifle, some pulleys and cable. We shot holes in the board to string cables through and used it as a chairlift to cross the river. I was quickly elected to wade the icy waters one last time to string the cable.

Morris returned, saw the contraption and sarcastically inquired, "What in the *Hell* is this?" Nevertheless, he got one leg over the board and rode it across. We used it from then on but Morris always complained that the chairlift wasn't big enough for both of his legs.

Now that we were no longer wading to Wade Quarry, we could have changed the quarry name to reflect the mode of access. "Bitching While Riding Across One Legged Quarry" was just too unwieldy.

When it's Dinner Time in Boneville

(Sing to the tune *When It's Roundup Time in Texas*)

When it's dinnertime in Boneville and I'm ridin' Lulu Belle
Then I long to be in Boneville, even though it's hot as hell!
Just to smell the gullion fryin' while it's sizzlin' in the pan
And to walk into a stinking tent and eat it like a man.
Just to see the scraper diggin' while I'm ringin' on the gong.
And to hear the Chiefy bitchin' that the outfit's set up wrong.
How it beckons and I reckon I would work for any stake
Just to see again and be again back in Boneville on the Snake.

Chapter 10

Memories of My Youth
By Tom Lucas[19]

I first met Morris Skinner on the sidewalk in front of the high school in Ainsworth, Nebraska. I don't know how he identified me, but he introduced himself and offered me a job digging bones. I accepted at once because he was paying more than anyone else was—six dollars a day plus board. I had never heard of Morris before, but all of my friends had. They thought I was lucky to have been so chosen.

On the appointed day, he made sure that I had appropriate clothing. He also outfitted me with a "fart sack"[20] and we were off to Sioux County, a long trip for me. On the way, he pointed out geological formations. He told me tales of camp. We stopped on route at a road cut and Morris collected a camel skull. I was impressed but unable to see the skull even after he pointed out the details.

---

[19] Tom Lucas was born in 1924, in Missouri Valley, Iowa. His father was a county Judge in Central City, Nebraska. Both of Tom's parents were killed in a train/automobile accident when he was in the eighth grade. Tom came to Ainsworth, Nebraska to live with his sister Betty and attend high school. Tom Lucas is a practicing Psychiatrist.

[20] "Fart Sack" was the common name for a sleeping bag.

Camp was a tarpaper shack on a treeless prairie encircled by a barbed wire fence to keep out the cattle. A roofless, three-sided privy stood in back. Down the draw, just out of sight, was a second shack, the bunkhouse. Later Morris moved his homemade trailer house to a hill still further down the draw. The buildings were not in any particular orientation to the world or to each other. They had wooden plank floors full of knotholes. Walking on them would sometimes arouse a rattlesnake, which would produce a frightening buzz under the floor.

A herd of antelope lived in the neighborhood. Each time we drove down the road they would race along side until, at the last instant, they would all cross the road in front of us. When they had safely crossed over, they would stop and ignore us until we again came down the road. Roads were two wheel tracks worn into the prairie and they meandered over the hills following the easiest path. There were few fences.

Two dogs were part of the crew: Sapa Mahto Chescheela[21] and Pikey. They were black cocker spaniel brothers. Mahto was larger and stronger, bold and independent. He was cooperative, but on his own terms. He was the leader. Pikey was timid and shy. They kept the rest of the crew entertained.

When possible they both rode with their noses out the window. Morris usually let them out of the pickup some distance from camp. They would run down the trail sticking their noses into every kangaroo-mouse hole along the way. One time a cow on the ranch was struck dead by lightning. Morris took a hind leg from her and stashed it over the hill where the dogs would rush to several times a day. They would then return, uncomfortably full, to lie in the shade and burp.

Eventually Pikey was bitten on the nose by a rattlesnake that hid in one of the kangaroo mouse holes. Despite loving care by Morris, Pikey ended up quite debilitated and had to be retired to city life.

---

[21] "Sapa Mahto Chescheela" is Lakota Sioux for Little Black Bear. The order of words translates to "black bear little." He was called "Mahto" for short.

We dug for fossils in quarries in the sides of the hills where sand and gravel had been deposited in ancient riverbeds. Some quarries were large, extending for a hundred feet or more with from ten to thirty feet of height. Others were quite small. Fossils lay generally in the gravel layers in sand, but they were often in hard clay at the very bottom of the old channels. Before we could get to them, we first had to remove the soil on top. The bones themselves were then carefully uncovered with steel awls and hooks, whiskbrooms and small paintbrushes. Digging was usually easy, but a couple of the quarries contained partially consolidated material that required the use of pointed geology hammers.

Some bones were tough and strong. These we lifted out of the matrix and carried to a central location. They were dried and then saturated with shellac dissolved in alcohol. At the end of the day, we would cover them with a paper padding and then with a plaster of Paris cast to protect them during shipment. Often the bones were fragile and these required preparation in situ.

The finding of a fossil bone is equivalent to finding a gold nugget or a treasure chest. It never loses its thrill. It may be like fools' gold or an empty box and only a worthless chip. There is always the possibility that it is a skull or even the first bone of a complete skeleton of a previously unknown species of beast. I will have been the first to see it and for a short time, it will be all mine.

In Sioux County quarries, the distinct ring of the hook striking bone signaled the find to others. At first, I was glad to receive help. However, after gaining some knowledge and skill I often protected my discovery for as long as possible. In spite of my attempts at secrecy, it seemed that everyone in the area would hear the characteristic ping of hook on bone. In answer to the question, "What have you got?" I would give the standard answers, "Just a rib." or "Just a metatarsal." This seldom fooled Morris for long. Any unusual activity or sound would bring him to help. I was just as curious about others' discoveries and we all shared in the joy. Although we actively competed for the best locations, we never admitted this to each other.

Morris could identify a bone with only a couple of inches exposed. I quickly learned not to bet with him, but we used it like a game. We would all guess what a partially exposed fossil would turn out to be. The winner gained prestige points. Morris astounded us with his knowledge.

We moved large quantities of dirt in the quarries using short handled spades. We shoveled while sitting or while on our knees, moving the dirt behind us as we worked. Eventually these piles of dirt behind us had to be disposed of. In addition, we couldn't allow the face of the quarry to get too steep or undercut because of the danger of its caving in. We also had to remove this overburden periodically.

We used an old-fashioned scraper that was designed to be pulled by horses. Morris had rented a team and we kept them at the ranch house. We never used them, but instead pulled the scraper with a Model-A pickup that Morris had modified for such use.

I never knew what to call the car. Morris said her name was "Maudie," but he usually referred to her as "Lulu Belle" and occasionally as the "Whoopee." She had a canvas top, no cover over the engine, and two transmissions, one directly behind the other and connected with a short driveshaft.

To move dirt we hooked the scraper to the front of the car by a long cable and backed up. Sometimes because of hills and gullies, we used a pulley on a post so that the car could move in a different direction and still pull the scraper forward.

We attached another cable to the rear of the scraper. This cable was run through a series of pulleys attached to posts and then fastened to the rear of Maudie. As Maudie went forward, the scraper was pulled backward. This setup was usually triangular with Maudie, the scraper, and the posts forming the corners. In a long quarry, several posts were set up behind the quarry so that the return pulley could be moved from one to the other. The angle of return of the scraper was thus varied to return it to different points along the quarry.

Three men were ordinarily involved in using this setup. A loader guided the scraper into the dirt and filled it as the driver backed away. The scraper was pulled out onto the pile or, if possible, over a gully. Another man tipped it up and dumped it. The driver then drove forward which reversed the direction of the scraper. As the scraper returned to the back of the quarry, it had to be picked up manually and moved to the proper spot for loading again. The speed of the car and the action of the loader required close cooperation. In some setups, the driver could not see the loader and depended on hand signals from the dumper.

It sounds complicated, but in practice, it worked well, except for a few snags! The setup required some slack in the cable to accommodate movement of the scraper off-line and to allow the posts at the back of the quarry to be higher than the floor of the quarry. Occasionally the car would get off-line and take up the slack. This would jerk the heavy scraper toward the middle and even up off of the ground to hang in the air. This would occur suddenly when the car accidentally ran over the cable.

Some members of the crew preferred to drive Maudie. I didn't. Morris liked to do the loading. He had invented the system, he understood it, and he was the boss. He stretched it to the limit. He insisted the driver go as fast as possible. He never complained of being tired. He wrestled with the scraper, treating it as an opponent. He cursed. He sweat. He seemed to take it as a personal insult if he had to move a pulley to make it easier for himself. After the work was done, he was relaxed and content. On the occasions when he let me load the scraper, he always drove slowly and gently.

All of the system was subject to breakage, especially Maudie. She was not engineered for the heavy loads that we gave her. We broke axles, transmissions and drive shafts so often that we became adept at repairs.

We could even change a transmission in two or three hours. We visited junkyards wherever we traveled and bought up parts in anticipation of need. Morris once found a cattle gate made of Model-A axles. He traded the owner for a different brand.

We shipped fossils once a month. This was a festive time. We got to review our previous triumphs. Each specimen was numbered, recorded and packed for shipment. We built shipping crates according to our needs. We sent most to the American Museum in New York City by freight. We expressed the special study items that Mr. Frick requested to his estate on Long Island.

We often cheated a little by saving some specimens back when we had been particularly fortunate. This made our lean months look better. We tried to make monthly shipments average out in numbers of specimens and weight.

We made a trip to town each week for provisions and recreation. I usually would have preferred to stay and work. Work, to me, was recreation.

When school started in the fall, everyone left camp except for Morris and me. When camp was full, Morris was concerned with keeping everyone working. His occupation was bone digging. When there were only two of us, he returned to his profession of geology and paleontology. He mapped all of the quarry locations and correlated them with elevations. In doing so, he plotted the courses of the ancient channels. We prospected for new locations and worked the smaller quarries.

With only the two of us, Morris tended to think out loud. I learned a great deal both about geology and about him.

Morris was generous. He helped a former employee who was in trouble and needed money. After I no longer worked for him, he lent me a large sum of money. He gave it to me in cash from his safety deposit box and refused a note or interest. His only condition was that if something happened to him, I should give the money to Marie.

I worked for Morris for four seasons. My employment began in 1942 before I went into the military during World War II. I then worked in 1946 and 1947 after I returned from the service. In 1950, after I finished college and before I entered medical school, I again worked three months.

I have always been impressed by Morris' drive, his stamina, his strength and his scientific reasoning. I regretted his getting old almost as much as I regret my own aging. I can only estimate his influence on my own life. I wear a mustache. I am a gypsy looking forward to my next trip. I do everything I can with my own hands; I am reluctant to hire anything done for me. I still look for fossils.

Chapter 11

Digging Bones with Morris and Marie
By John L. Beattie[22]

In the summer of 1943, I had the privilege of working for Morris and Marie Skinner as a bone digger. Morris had also hired my cousin, William J. Lear, for the summer. I had been told that my Grandmother, Effie Lear, was a midwife who assisted at Morris' birth. My mother could always remember Morris' birthday because her beloved Gordon Setter, over which she had shot many a prairie chicken, had pups the same day. She loved to tell Morris the story about this memorable occurrence. She stressed, that in *her* mind, maybe the birth of the pups was more important! This was always a joke with them. I only tell this story to demonstrate that our families go back for almost a century. All of this started in the town of Springview, Nebraska in Keya Paha County.

My first week of employment with Morris consisted of a survey of an area of southwest Nebraska, close to the Kansas border. They were planning a large reservoir to cover this area. I believe that this is now the Harlan County Reservoir. The land that we did stratigraphy studies on is now covered with a large lake.

---

22 John L. Beattie died in October of 1999.

I got some idea of what the summer would be like because we shot and ate numerous cottontail rabbits. I remember going into a small town in Colby, Kansas. We bought the last .22-caliber shells that the town had in stock. As munitions were in short supply due to World War II, Morris thought this was a wonderful find. These shells supplied much of our food during the rest of the summer.[23] I was very interested in visiting Colby because my father, Dr. Leo Beattie, was born there. My Grandfather, Jasper Beattie, M.D., had practiced in Colby for a number of years. Colby was the last town ever to see a trail herd brought in from Texas. This was the end of the railroad at that time. The herds from Texas were then shipped to the East for slaughter.

After several weeks of surveying and studying the stratigraphy at the reservoir site, we returned to Ainsworth. We loaded all of our digging supplies into the pick-up truck, we attached "Old Maudie" and we towed her to Western Nebraska. "Old Maudie" was a 1928 Ford with two transmissions obtained from junkyards. Morris had converted her into an excavation tool for the fossil digging at our quarry sites. We spent one memorable day prior to our departure in the Witt junkyard harvesting old transmissions from late 1928 and 1930 Model A Ford trucks. "Old Maudie" shucked transmissions like Iowa farmers shuck corn.

The trip to Western Nebraska was uneventful. We headquartered in the line shack on a ranch owned by Clarence Kilpatrick. He and his neighbors gave Morris and his bone diggers access to their valuable grazing lands as that is where most of the fossil quarries were. A large water storage tank was present at the line shack. Each evening we delighted in diving from the windmill tower into this large tank for our evening bath and recreation. Fritz Skinner, Morris' son, was another

---

[23] Morris was an uncanny shot. He had a Colt Woodsman .22 automatic pistol and any animal or bird was in peril if Morris deemed it dinner. Game fell wherever he pointed it.

member of the digging party. Marie, Morris and their daughter Barbara luxuriated in the line shack while we lonely bone diggers, Bill, Fritz and I, were relegated to sleeping bags in a makeshift bunkhouse separate from the line shack.

I have several distinct memories about the happenings on the Kilpatrick ranch. Morris killed a rattlesnake close to the line shack one morning. That evening when I crawled into my sleeping bag, I encountered a rattlesnake. Fortunately, this rattlesnake was deceased. Young Fritz Skinner had placed it in my sleeping bag. After experiencing a near cardiac arrest, I finally retrieved my senses and lusted for revenge. Had it not been for my love of Morris and Marie, I am afraid that dear Fritz might have been on his way to Heaven.

Another very impressive memory of the Kilpatrick ranch was visiting the ranch house where Mr. Kilpatrick lived. Numerous peacocks strutted around the ranch. This was the first time I had ever had close association with this beautiful bird. I understand that Clarence was divorced from his wife and had a beautiful Hispanic mistress. You can understand the effect this beautiful lady had on our 16-year old hormones. Bill and I were very envious of Clarence Kilpatrick.

I had the privilege to meet Mr. Kilpatrick again in 1956 as I returned from my three-month tour at Sun Valley where I had worked treating ski fractures. We spent about four hours on the train talking about my adventures and discussing Morris Skinner and his numerous accomplishments.

Another distinct memory about living in the line shack is that we were only several miles from the center of a practice bombing range used by the Scottsbluff Air Force Base. These B-17s and B-24s used practice bombs containing an incendiary powder that would ignite on impact. This let the bombardier check the accuracy of his bombing run. Unfortunately, the incendiary flash would ignite many prairie fires. We spent a number of days helping ranchers put out the fires. Morris' wisdom about prairie fires and the advice he gave me probably saved my life on one occasion. We would fight the fires with wet gunnysacks. Morris told me that if the wind ever changed, I should wrap the gunny-sack around me and run through the blaze to the burned-out side. The fire singed my hair and scorched my clothing but I did not suffer any serious injuries.

The real physical labor part of the bone digging operation was the excavation to uncover fossils. A triangular set of cables and a scraper pulled by "Old Maudie" moved the dirt. It became obvious after several weeks of digging that the football fullback (me) was needed for the physical effort, while my less ambitious and more cunning, handsome cousin was allowed to drive "Maudie." I never fully recovered from knowing how much better Bill had it driving than I did wrestling the scraper.

We also had many interesting expeditions searching for fossils in out-croppings. I remember one area east of Agate, Nebraska on the Red Morava Ranch where we found many *Stenomylus* camel skeletons. We took several weeks to excavate these finds. A number of years ago, while visiting the Field Museum in Chicago, I saw this type of camel dis-played. I presume they were donated or on loan from the American Museum of Natural History. One of Morris' very close friends, a Mr. Jim Quinn[24], who was also a paleontologist with Ainsworth roots, worked with this museum.

I must say that Morris was a hard taskmaster. He was very opinion-ated about what we should do. I did my best to obey him as far as my physical abilities were concerned. I must say that I have never met a more physically tough or mentally tough man than Morris Skinner. He was extremely intelligent. I understand that he had dyslexia and I think this may have accounted for his excellent memory and his abilities to make many decisions. My association with the Skinners helped me as I traveled through life, particularly when one considers the rigors of

---

[24] See "James Harrison Quinn (1906-1977)" in this publication.

World War II, the Korean War, the indignities of medical school, residency and the problems of 48 years of medical practice. During my internship in Charleston, South Carolina, I wrote Morris and Marie about my being stationed there. They introduced me to a former associate who did much for me as far as my social life was concerned and gave me an insight into the life of old Charleston.

The culmination of these experiences came to fruition at the commemoration of the Burge Quarry Site in Cherry County in 1990. I spoke as a representative of the bone diggers. During this talk, I expressed my love and affection for Morris and particularly Marie. I had a real schoolboy crush on Marie. I still have it and my wife doesn't mind.—(John L. Beattie, M.D., F.A.C.S.)

Chapter 12

Memories of Maudie
By William J. Lear[25]

During the summer of 1943, Morris Skinner asked my cousin, John Beattie, and me to work for him digging bones for the American Museum of Natural History. This invitation was, more or less, a result of the war situation. At that time, no civilians who were high school graduates could be involved in other than military jobs or those important to the war effort. John was a physically mature seventeen-year-old of moderate stature and good development. I was sixteen, tall, adolescent and gangly with incomplete physical development. That was probably my outstanding characteristic.

I first met Maudie in early June of 1943 when I wandered into a shed at the Skinner farm on the edge of Ainsworth, Nebraska. Maudie was a very unimpressive sight. She was a well-worn specimen of a Model-A pickup in sore need of cosmetic surgery. She was rough in the extreme and showed evidence of hard use. Surrounding her was equipment, the use of which was mysterious at best. There were also plaster casts of lumps of dirt, which did not clarify her use in any way.

---

[25]William J. "Bill" Lear's father, William D. Lear was the doctor, in Ainsworth, Nebraska. William J. Lear, the author of this story, also practices medicine.

We spent the first two weeks, or so, of our employment with Morris helping him do maintenance on machinery and equipment while preparing for the summer's pursuit of fossils. We took breaks to prospect the canyons around Ainsworth. We also made a trip to southern Nebraska, to the loess hills near Trenton and the Kansas border. At the conclusion of this trip, we returned for a weekend to Ainsworth and thereafter set out for Sioux County north of Scottsbluff, Nebraska. That is where we spent most of the remainder of the summer.

As previously stated, I had no idea how Maudie was useful in the mining of fossils. Much to my surprise, she was included in our short wagon train to Sioux County. In fact, Morris towed her ignominiously behind his primary mode of transportation, a 1938 Ford pickup.

We arrived without incident in Sioux County in either late June or early July and set up camp. We established residence in a rather rustic way and began to pursue our primary occupation: the discovery, mining, and packaging of bones.

Maudie just sat there while Morris led us on scouting expeditions into the cut-banks and draws. This was a significant part of the surface makeup of Sioux County. The productive method for producing bones in quantities is to isolate an area where fossils are in good supply and then mine them. We directed our initial efforts then to finding an area worthy of development. After determining the potential of several areas, we set up to dig bones. This is not a complicated procedure but it is very labor intensive.

At last, Maudie came to the fore as an important participant in our efforts. She was considerably more than met the eye. The Model-A motor was an efficient, albeit not all too powerful, 4-cylinder unit. Morris kept Maudie in sharp tune and she was a reliable workhorse. Since she was lacking in power, Morris found it necessary to beef-up her drive train. Morris asked different mechanics to insert an additional, tandem transmission. The local consultants felt that this was not a possibility and predicted that it would not work. This was the challenge that Morris needed. He installed the second transmission himself. With the modification, we could now select an assortment of gears such as double low, an unstoppable combination.

Maudie also had squat wheels with diminished circumference as compared to the wheels of a standard Model-A. The rims were fitted with substantially wider tires so that the additional mechanical advantage would produce more traction. Maudie's importance was simply that we needed to remove a fair amount of earth from above the layers that contained fossils and she performed this task. Maudie was the power unit and we connected her by cables to a piece of equipment that looks like the front end of a large scoop shovel. The scoop had a handle on both sides and a harness to attach it to the cables. We needed power not only to fill the scoop but also to return the scoop to the next load of earth. We fastened this return line to pulleys on three or four deeply anchored posts. We could route the return cable to the post that provided the best direction for the travel of the scoop. After placement of the anchor poles and making of adjustments with the cables and pulleys in relationship to Maudie, the removal of overlying dirt proceeded quite smoothly.

For the best efficiency, the process required three people. One was responsible for the direction and loading of the scoop, one to shepherd and empty the scoop and one to drive Maudie.

John Beattie or I would empty the slip or drive Maudie. John was a much more physical person than I was then, and for that matter, was throughout his life. His approach to a problem was direct, forceful, and immediate without too much attention to finesse. My approach has always been hesitant and cautious, slow and easy, without the incorporation of sudden moves. It soon became apparent that my personality was more appropriate for driving Maudie than was John's. This suited me fine. Of the two chores, emptying the scoop was substantially harder work. Driving Maudie required obeying hand signals given by the other two operators to move the scoop gently and powerfully through its required range of motion. Translating the hand signals to the application of force, with some restraint and delicacy, was the required skill.

Because of my reluctance to do hard work, I managed to develop this skill to the point where I spent a good share of my bone digging days driving this old, but competent and well-loved vehicle named Maudie.

Chapter 13

Bones of the Fifties
By Loren M. Toohey

We were about as efficient as three people could be. That, to me, expresses my impression of fieldwork in the 1950's. The party consisted of Morris, Fritz and me or Morris, Bob Lamb and me because the makeup of the group depended upon the time of the year.

Morris had the Ford pickup modified with sideboards to hold food, stove, gasoline cans and field supplies. We carried two cream cans for water. A large metal lard canister contained plaster. Burlap, sleeping bags, mosquito netting and air mattresses went somewhere—clean clothing in a cardboard box—a wooden box for dishes—coffee pot and pans were all neatly stowed in the box or sideboards. An L-shaped piece of metal welded to the fender and box of the pickup held two one-gallon jugs neatly wrapped in burlap. Yes, at one time Coca-Cola syrup came in one-gallon glass jugs. I can't remember if we had two gallons on each side of the truck. Somewhere we had three, somewhat modified, short-handle shovels. In the sideboards, we had an umbrella tent for an emergency. Geology hammers, backpacks, paper tape, newspaper, and awls rode somewhere in a convenient place.

We had a minimum amount of stuff and could move easily and rapidly; so off we would go for two weeks, ready to challenge the elements and outcrops.

The menu was simple and adequate. Breakfast consisted of two eggs each and coffee for one morning followed by pancakes for the next. Lunch was sardines, crackers and an orange (no orange for Morris) or hard salami, crackers and an orange. Supper was usually a half-pound of hamburger each and coffee. At my rather advanced age, I still like hamburger.

When we left Ainsworth, we had packed packages of frozen hamburger to last about a week into shredded paper inside a cardboard box. We might leave a package out to thaw and therefore be ready for supper that evening. Usually, depending upon the daily temperatures, the frozen packages would be edible to the last. Sometimes the coyotes would get a taste of beef.

Water was always scarce and we didn't waste any. We didn't shave or bathe and we washed our face with a cup of water. We needed most of the water we carried for coffee, washing dishes, drinking and casting. It seemed that windmills were almost non-existent. I remember taking a bath and washing my hair in Pumpkin Creek. I believe I was cleaner before I got into that silt-laden creek!

When supplies ran low, we would head for the nearest town for gas and food. One time when we camped in one of the Dakotas, the field corn hit its peak for human consumption; so, we gorged ourselves for a few days. It didn't take long to fix either one of the two meals.

If we were doing one-nighters, we usually selected a campsite before the sun went down and in a place where mosquitoes might be few in numbers. We pitched camp in the dark several times, but we never woke up to a rancher staring at us. When we had selected a campsite, Fritz or Bob and I would grab three shovels, bedrolls, air mattresses, tarps and mosquito nets. Morris got out the box of dishes, water, food and the stove.

On the side of the pickup were tie-downs or hooks. On the ground below these, we laid a tarp or two upon which we unrolled the air mattresses. A pump Morris found somewhere was screwed into one of the sparkplug holes and a hose on the other end screwed onto the valve of the mattress. Presto! Start the engine and air from the exhaust valve would soon fill the mattress. We repeated the procedure twice more and bedrolls were ready to be unrolled. Two ropes, two shovels and the side of the pickup made a convenient square for attaching the mosquito nets. Tucking the sides and ends of the netting well under each mattress kept all the varmints out.

While all of this was going on Morris had started the camp stove, was cooking the hamburger, the coffee was beginning to perk, and a pan of water was ready to be heated as soon as a burner was free. We always attempted to keep plates and dishes to a minimum. You would be surprised to know how minimum, minimum could really be. Shortly after we downed the last drop of coffee, we washed and dried dishes and got things ready for breakfast. Usually the mosquitoes made their appearance about this time. After urinating and getting a fix on the cacti, it was time to undress and get into bed as quickly as possible. It was a great personal satisfaction to be inside while the mosquitoes were buzzing and trying their best to get at the fresh meat only inches away. The still of the night was broken by a coyote now and then, stars were a dime a dozen, and a full moon was something to witness. Did we put our clothes and shoes where they would not get wet if it rained?

Morris always made enough noise the following morning passing wind, hitting the truck with a shoe, slamming a door, to prevent further dozing. Unscrew the valve on the air mattress, dress, and do everything in the reverse order while Morris prepared breakfast. God bless the inventor of air mattresses and mosquito nets!

Get your backpack. Make sure it is loaded with awls, notebook, indelible pencil, newspaper and paper tape. If you let your roll of paper tape get wet, you were in trouble. Toss in your lunch of an orange, crackers and sardines. Rusty sardine cans are now our artifacts left to society. Grab your hat and geology hammer and fill your belly with water. Your next drink of water would be when you returned to the truck!

One day when the mercury went to the top of the thermometer, we had full packs of fossils and a mile or so to walk uphill to the truck. I do believe the bloody truck had moved itself two miles further during our prospecting! I don't recall ever being so parched. Fritz looked like warmed-over death. Morris wouldn't admit a thing. The remaining water in the jugs, when we finally got to the top of the hill, was the temperature of warm pee. Five miles of pasture, gullies, and ten thousand gates of barbed wire separated us from a cold beer in Oelrichs, South Dakota. No, we didn't crawl up the steps to the bar but we sort of wobbled.

Why did we go where we did? I don't know if Morris submitted a summer schedule to Mr. Frick, or if directions were in the reverse order or both. During this period, we worked in Cherry, Brown, Dawes and Sheridan counties and the North Platte River Valley in Nebraska, many areas of South Dakota, southwestern North Dakota and the eastern part of Montana.

I have read what few field notes I still have from those days and am amazed that I have forgotten so much about so many important events. It is as though someone had pulled a cord and flushed half of my memory somewhere into outer space—if that is where your memory disappears. How could I forget the unusual titanothere skull[26] from the east end of Capitol Rock? How could I forget the creodont mandible from what we now call the Rocky Ford Ash? How could I forget…? I won't summarize our efforts because no one cares about what we found at

---

[26] We shipped the unusual titanothere skull to New York in 1954. Thirty-five years later, its nap was disturbed in the dark vault of the American Museum when preparators opened the cast. The scientific community was apprised of this unique specimen in 1995. (Loren M. Toohey, 15 Jun 1999)

Pass Creek, White Buttes, a prospect on the Snake River that developed into a good quarry, etc.

I think we were the first to collect in some areas. One place required us to crawl on hands and knees along a knife-like ridge, which had a rubble-strewn steep slope followed by a vertical drop of 100-150 feet. When we could no longer crawl any further, we sat down and inched our way along. One false move and we would have been broken or dead. Anyway, we found the pot of gold once we arrived at a long narrow flat area. Morris referred to this place on Sheep Mountain as Split-Ass Ridge.

We went into an area, collected the small material and blocked out the larger specimens for casting. The following day we might cast and collect those we had prepared or we might return in two or three weeks, on our way to Ainsworth, and do the casting. A good sharp point on a modified geology hammer and the correct movements with your wrist would let you get the smaller skulls and jaws into a nice sturdy block. Wrap them tightly in newspaper and apply the brown paper tape. Write the necessary information on the dampened upper surface of the tape. The best and easiest method is to spit and rub the tape and put the indelible pencil to work.

I often wondered what the inventor of paper tape had against us. The taste of glue lasted all day and it sucked the last drop of saliva from your mouth. People whose salivary glands functioned poorly did not last long in the field. The purple tint of the writing is legible after many, many years. With this same indelible pencil, we wrote on the plaster of Paris casts while they were drying. I wonder if indelible pencils are still being manufactured or if you ask for one, is it like asking for kitchen matches?

If we collected enough fossils the first week out and the truck was getting too full, we could make a trip to the nearest town and find some lumber, straw and nails. We would make a few boxes and ship them from there. Our usual procedure, however, was to hit the field for two weeks and then return to Ainsworth for packing and shipping. When lumber prices went up, Morris decided we could make boxes from discarded crates we found in the alleys of Ainsworth. We made some pretty good, nonstandard sized boxes.

We packed the carnivores and camels separately, if we had enough. We labeled the fossils and samples of volcanic ash with the information relating to the sections we had measured. This same information made its way into the packing lists diligently typed by Marie. As soon as we had shipped the boxes, it was time to prepare for another two weeks in the field. I have always regretted never seeing any of the specimens we collected after they reached the museum.

We measured the vertical sections of the outcrops to tie the specimens to the stratigraphy. Sometimes Morris might come back later and measure a section if we had been pushed for time. These section books became as valuable as the gold bars in Fort Knox. Today they form a most important part of the documentation relating to the specimens that we collected.

We saw a lot of rocks and beautiful country. If you haven't looked down from Porcupine Butte or Eagle Nest Butte, you should. Go up to the top of Blue Butte, Killdeer Mountains, or Long Pine Hills. Look down from the edge of North Slim Buttes and spot a lone bull elk. Find some wild strawberries for the first time. Two deer might walk close to camp knowing they are safe. You might share a watermelon with a young Indian boy and his sister.

Each day was a new experience. At White Butte, I found a Mississippian coral in the channel that produced the *anthracothere* partial skull. The top of one butte was littered with granite boulders. A vertical clastic dike, some twelve feet wide, contained rounded rocks up to 12 inches in diameter. These things cause me to wonder how and why. A bed of sedimentary quartzite, three-feet thick, also caused some head scratching. We saw all sorts of sedimentary structures, facies and bedding. The fossils in the Whitney-like rocks at North Slim Buttes were "hot" whereas the fossils in the lower and Orella-like rocks were not radioactive. All of our trips were successful. What an education! It also did not take long to learn which sage was the useful one!

Bits of humor and excitement broke the routine. I remember the time when, for some unknown reason, Morris stuck his head over the plaster bucket at the exact moment I was trying to get the plaster settled. I dropped the bucket a bit too hard and the powdery mess exploded upward into Morris' face, hair, eyebrows, and mustache. An unearthly calm settled over the Skinner estate. It became so quiet you could have heard a mouse fart in the corncrib. Morris walked away without a word. Fritz and I breathed a sigh of relief.

Imagine, if you can, Fritz and me directing left-handed Morris how to back down a steep, very narrow logging trail. He kept turning the pickup the direction opposite to our instructions.

When resurfacing roads out West, the contractors piled gravel ridges a foot or two in height along the shoulder of the road. Imagine driving fifty miles-per-hour over a slight rise after sundown and finding a long pile of gravel directly in front of you. Morris turned left and then back right with half of the truck now in the ditch. He got things under control and brought us to a safe stop. We got out and stood there for several minutes cussing the gravel, the contractor, the county, the state—cussing anything else that came to mind.

We were baked, almost frozen, and not quite drowned. We cooled off, warmed up and dried out.
—Loren Toohey, 7 Dec 1992

Chapter 14

Fine Company

Whenever acquaintances gather to reminisce about the days of yore, you might hear a familiar phrase, "We sure were in fine company back then!" This chapter is about the fine company of other associates of Morris Skinner. Some declined or were unable to contribute personal remembrances. Then there are others with whom we have lost touch.

Leonard Nelson worked for Morris before and after World War II. Leonard both prospected with Ralph Mefferd and did quarry work in Sioux County, Nebraska with Morris, Tom Lucas, *Bill Laverty* and Fritz Skinner. Leonard first went into pre-med at the University of Nebraska but then switched to geology. He became a very competent geologist and oilman. Marie recalls, "It was Leonard Nelson who said that he didn't need to be toughened up by the army after working with Morris."

Marie continues, "There were others who did very well in the field with Morris. *Alan Lamb*, daughter Barbara's ex-husband, went into oil geology because of his interest in the fieldwork and he is a successful oilman in Oklahoma City. His brother, *Bob Lamb*, also worked and did well—always complaining that Morris seemed to know when he had been out the night before 'tomcatting' because he would take him to the steepest canyons for prospecting the next day. Even young *Ralph Mefferd Jr.* did yeomanship with the crew and still speaks fondly of it. He loved cattle and horses and that is his occupation."

Haakon Dehlin, a Norwegian, was Mr. Frick's assistant on mastodon research. He was well educated, very handsome, and good in both research and collecting. He had been an Olympic diver in the 1920s. He joined Morris' field crew for two seasons in the early thirties, but his drinking nearly wrecked the reputations of local bone diggers by association. Don Barton, then a Meadville teenager, remembers Haakon as a sarcastic person who spoke very poor English but who spent money like there was no Depression. Hence, people catered to him who would not have otherwise. Don, a smart-aleck kid, would take none of his guff and that impressed Haakon. He seemed to admire the trait of someone talking back to him. They struck up a friendship and Haakon volunteered to teach Don to dive. One summer, in a hidden lagoon along the Niobrara River called the Meadville Bayou, Don Barton perfected his swan dive with the help of an Olympian who was far from his homeland. Those who drank or dove with Haakon Dehlin may remember him as fine company, but he was not suited to the disciplines that Morris required. Morris refused to accept him in the field after those two years.

Dale Kirkpatrick worked for two seasons. He will never forget that when a quarry wall fell and covered him to the neck, Morris asked, "You alright?" When Dale nodded, "I guess so," Morris offered him a shovel and ordered, "Well, then, dig yourself out."

Carl Elfgrin was a contemporary of my brother, Bob Emry. Carl and his family came from Connecticut to Nebraska in a horse-drawn covered wagon, gaining modest fame along the way. It took two summer seasons to complete the long trip. The Elfgrins were a large, young and adventuresome family. Upon arrival in Nebraska, Carl's father, Ernie, opened a machine shop in Berwyn. Carl loved to hunt and fish and to relive the old west and its ways. Someone with so much experience in the great outdoors, and at such an early age, was a made-to-order bone-digging apprentice. Camp food was no problem for Carl, so long as there was plenty of it. Prospecting the wilds was Carl's dream come true. He was a dedicated bone digger and an asset to any party. His leaving the bone digging crew opened a spot for me. Morris hired me as Carl's replacement. Carl continued to be an outdoor enthusiast and was renowned throughout the state for his ability to make arrowheads and shoot muskets.

Photo 9: Pumpkin Buttes, Wyoming—Kenny Weichelman as "scale"
—July 1962

Kenneth Weichelman, the son of the Ainsworth shoe-repairman, was a contemporary of mine. He worked about four summers in the 1960s. He was a reliable worker who attended the University of Nebraska during the school year. Marie recalls that Kenny could "run like a deer." I don't remember his running ability, but I do know we were all proficient at walking for miles on a tall drink of water. When I tried to recall Kenny's fleet feet, I remembered that Morris always liked "scale" in his photos. A person near the badland exposures that Morris was about to photograph would add that ingredient. I can now clearly remember occasions when Morris would send Kenny into the far off landscapes to fulfill that task. We might be on one butte looking at another in the distance. "Hey! Kenny? How about running over to the north butte— climb up there near the cap-rock…" The last phrase shouldn't be in quotes but I recall several running requests of Kenny that were similar. Kenny may have run there and back. Unfortunately, I excelled at being a scale a few times too. Nonetheless, when Kenny was around, and Morris got out his "4 by 5" camera that he used for high quality black and white photos, I sensed what was coming. I would grab the light meter and read the lumens from every possible aspect, to look busy, in hopes that when Morris spoke, he would begin with "Hey! Kenny?" When and if Morris spoke, he would certainly have to interrupt my running dialog, "The shadow there on the north butte is a 100th at f16, then below the cloud I get a 200th at f22, then…" I also mention Kenny Weichelman in my recollections, *Part 3—Bone Digging Days*, which follows.

Dan Chaney, employed by the Department of Paleobiology at the National Museum of Natural History for many years, also spent earlier collecting seasons with either Ted Galusha or Morris. Ted also led field expeditions for the Frick Laboratory.

Paul Sandaar never worked for the Frick lab but he spent one summer with us when I was on the crew. I believe he was a professor from the University of Amsterdam, Holland. He was twenty-nine that summer but he looked younger. This was due, in part, to his lack of facial hair. His youthful face brought him some chagrin on the rare occasions when we stopped off for a cold brew. Bartenders sometimes asked for his ID. This scrutiny was probably not how he had envisioned the Wild West, where Billy the Kid sauntered up to the bar and was served. Paul was fluent in several languages and an expert at his trade. Paul and Morris talked geology at a level far beyond my understanding. Paul was always pleasant and friendly and had an ever-present twinkle in his blue eyes. He was fine company indeed.

One day, after prospecting the Sand Draw north of Ainsworth, we were driving back to town. A badger happened across the road as we were driving along. Paul had a movie camera and wanted to "shoot" the badger for his photographic trophy room. By the time Paul got his camera out, the badger had loped into a grassy meadow on a beeline in an easterly direction. Paul crawled through the fence and took chase with one eye in his viewfinder. When Paul ran to within about six feet of the badger, the badger sensed a violation of its personal space. The badger stopped, turned, snarled, and jumped at Paul. Paul stopped, turned, vocalized at a pitch several octaves higher than a snarl, and set an international record for the one hundred-yard dash back to the pickup. Meanwhile, the badger was setting a speed record of its own back on a beeline in an easterly direction. The separation rate between the two was awesome. For the few seconds of the encounter the badger was in fine company and didn't even know it.

Mr. Frick sent his two grandsons to the field separately. *Townsend Burden III* came west one summer and became part of our field crew in Wyoming, Nebraska and South Dakota. I also mention Townie in *Part 3—Bone Digging Days*. His younger brother, *Henry Burden*, came with a friend, *Connally Keating*, and worked most of the summer of 1966. I remember that summer well. I had graduated in June from Colorado State University. I was also a brand new USAF Second Lieutenant awaiting orders to active duty. It was my last summer as a bone digger. I welcomed Henry and Connally's arrival with little enthusiasm. I thought that, considering I was a veteran of six previous summers with Morris, a college graduate and a military officer as well, I had finally advanced to a position of seniority and entitled to the sacred perk of riding inside the pickup cab. I was wrong. I suffered a few slow burns riding along in the hot, windy pickup box as Henry and his young buddy sat up front with Morris.

Nonetheless, Henry and Connally were friendly guys and easy to like. I saw a marked difference in work ethic between older brother, Townie, and this younger pair. I always felt that Henry and Connally saw their summer as vacation and I suppose it was. I noted with some jealousy how easily they could spot a bit of shade and how long they could homestead in it.

Morris, I am sure to get us all out of his hair, sent the three of us packing to the Snake River. I was their boss. Morris made sure they understood that what I said went. They both respected my authority and we never had any problems except we never came home with many fossils. Fossils never seem to expose themselves in shady places.

I couldn't understand how Henry and Connally could sit and read or snooze while we were traveling through beautiful landscapes. I had my eyes out the window and head on a pivot wherever we went. When we were driving back from our expedition to the Snake River, they both had their noses in books or were napping by the time we arrived in Valentine, Nebraska. While they were so preoccupied, I took a side trip out to the Niobrara Game Refuge, turned into the buffalo pasture, and drove right up to a giant, bull buffalo resting in the roadway. When I stopped, Henry and Connally looked up and wondered why. As the impressive old bull clambered to his feet, I stated indifferently, "These dang things are always getting in the way." Henry and Connally stared wide-eyed and slack jawed as the buffalo slowly moved out of our way. I never mentioned that I had taken them sightseeing to a game refuge.

My flying logbook shows that I took Henry Burden and Connally Keating up for a spin on 7 August 1966. I don't remember where we flew or what we saw. They may not remember either, as it is highly probable that they were reading books or snoozing as we flew.

Wiley Lentz was never an employee of the Frick Laboratory nor did he go to the field with Morris. Even still, he had more interest in paleontology than did some others who had that advantage. Wiley was a meat-cutter in the Red and White grocery store in Ainsworth, Nebraska, and he hunted fossils for a hobby. Morris liked him very much and was proud of his knowledge and interest. He and Morris struck up a relationship that endured. Wiley drank in everything that Morris could teach him.

Because Wiley was a butcher, he had a formidable knowledge of beef bones. Wiley discovered the cardiac bone in a beef heart during his butchering research and showed Morris. Although the cardiac bone has all the characteristics of a true bone, Morris showed it to the bone experts at the museum and absolutely no one could guess what it was.

Morris liked to fool the experts. When Reid Macdonald (Paleontologist at the University of Idaho) and his wife, Laurie, came to visit, Morris asked Reid to identify the bone. Morris didn't know that Malcolm McKenna, at the American Museum of Natural History, had tipped off Reid about the cardiac bone. It was a way for Malcolm to get back at Morris for some previous ruse. Reid acted as if he had no idea what the bone was and handed it to his "non-expert" wife. She quickly examined it and immediately and with authority told everyone that she was quite certain it was a cardiac bone from a cow. That let all of the air out of Morris' sails. Marie said that Morris was so disappointed that she didn't think he ever tried to catch a specialist on the cardiac bone again. There were plenty of ways to play tricks on each other. Morris could do it, but Malcolm, who was a master of the art, sometimes bested him.

Morris F. Skinner Jr. known as Fritz grew up in his dad's bone digging camp. When he was old enough, he joined the crew and worked for nearly fifteen years. Fritz went to medical school and turned his knowledge of fossil bones into a career as an orthopedic surgeon. Fritz occasionally visited Ainsworth on vacation and went to the field with us.

A father/son relationship is fraught with uncertainties even in the perfect setting. I believe Morris was a tough but fair employer. I suppose he was also a tough but fair dad. Even still, there are some orders an orthopedic surgeon may find irksome and bordering on cruel and unusual. When Fritz was on vacation one summer, Morris put us to work. The old privy needed a new hole and Morris stated adamantly that he didn't want to move the "sally-house" ever again. The workers used the outhouse when we were working in Ainsworth. Morris did too, when he didn't want to trek to the house.

Digging an outhouse hole to me was not much different from digging for fossils. Paul Sandaar, ever pleasant, joined in our dig with little hesitation. I detected that Fritz, a professional in his own right and with a limited time at home, would have rather spent his vacation doing something more fulfilling than such a chore for his dad. However, when Morris told us to jump, we often asked "how high?" on the way up. You may remember Howard Williamson's observation; "A bone-digging husband who is mad at another guy's wife—justifiably or not—is a more productive member of the group." Similar human relationships were in play, at least I thought so, the day we dug the outhouse hole.

Fritz made the dirt fly and we hit ten feet before we took a break. By then humor had started to seep into the outhouse hole and we dug for the fun of it. We stopped at a depth that could not have been too much short of the water table. By then, we were taking the dirt out via a bucket on a rope. Morris came to check our progress. The new outhouse hole met with his approval. Morris said he believed "he could finish his business and leave before it ever hit bottom." Sometimes a father will praise a son in unconventional ways.

Fritz was the flag boy during the early years of his employment. He signaled when the scraper was loaded and emptied so the driver could reverse course. He was also in charge of loading the pickup for prospecting expeditions. When he inadvertently forgot something Fritz recalls, "Father never said a word—just indicated 'well that's it.'"

Morris told Marie that Fritz found the majority of fossils when they were in the field. Fritz enjoyed prospecting and was good at it. Morris prepared most of his son's finds, as Fritz didn't care as much about digging them out as finding them.

As Morris was very careful about nepotism, if someone had to ride in the back of the pickup it was Fritz. In fact, he didn't get to ride in the front seat until 1948. From Fritz' point of view, the policy may have smacked of reverse discrimination. If I could have labeled my feelings when Henry and Connally came to camp, I would have called it that. Of course, the term reverse discrimination did not exist in 1948 or even in 1966.

Before Fritz settled down to a long career in medicine in Southern California, he served in the Navy during the Vietnam War. He was stationed in Vietnam from February 1968 to February 1969. He was first assigned to Phu Bai and then to Quang Tri where he was Chief of Orthopedics. His recollections of life and death in the medical wards are grim. On his busiest shifts, he performed eight or nine operations. Fritz recently returned to his beloved Nebraska. He is continuing his medical practice in Ainsworth and Grand Island, Nebraska and in Gregory, South Dakota

Photo 10: Near Redington, Arizona—Robert Emry Listing Fossils—Jan 1962

My brother Bob, *Robert J. Emry,* began working for Morris in about 1958. He spent his first winter in Arizona under Ted Galusha and returned to Morris' field crew the following summer. The next winter, Bob worked with Morris near Clarendon, Texas. Bob continued to work the following summer of 1960 in Wyoming with Morris. That was the year when I became a bone digger.

Bob showed a lot of promise and before his twenty-first birthday, he was trusted to lead a field crew to Arizona for the winter of 1961-62. His crew consisted of two, including Bob. I was Bob's able assistant. We worked in fossil quarries near Redington (northeast of Tucson), Wikieup (southeast of Kingman), and Prescott. Bob was very conscientious. Each day was just as long and each specimen was treated with just as much care, as if we had someone more senior in command. Working with Bob was a good experience for me. I was on my own for the first winter of my life, but Bob was there to ease me out into the big old world.

It was a time when two grubby young fellows from Ainsworth, Nebraska could walk into a store in Tucson, Arizona and buy a case of dynamite, blasting caps, and fuses, no questions asked. I figured Brother Bob knew what he was doing so I welcomed any method that would quickly remove overburden from the quarries. Our tool supply was limited and our only machine was an aging Jeep that Maudie, of an earlier generation, would have put to shame. The layers of tough clay, above the fossils, didn't easily yield to manpower. Therefore, picks, shovels, sweat, and dynamite removed this overburden.

We had some steel rods, about seven feet long, with chisel bits on the ends, called churn-drills. If you lifted and dropped a heavy, churn-drill repeatedly in one spot, with a slight twist between drops, you would gradually gnaw a hole in the clay. After we churned for a while, we would pour water into the hole and make mud of the loosened clay. We removed the mud with a muck-stick, a one by one-inch square wooden shaft the length of a churn-drill. We repeatedly thrust it into the hole, mucked it around and pulled out whatever mud that would stick to its sides. Once we had removed the mud, we poured more water into the hole and again pumped iron. After numerous repetitions with the steel chisel, we had gnawed another foot or two. We repeated the process until we were finally on our knees with only a handhold of drill exposed above the hole. Our arms would be so tired that we could hardly lift them, let alone the churn-drill, which now also resisted movement because of the drag produced by the sides of the muddy hole.

One day, I stood to straighten my back and rest my arms. My hands had been wrapped around that churn-drill for hours, lifting and dropping, lifting and dropping. As I flexed my fingers, the skin of one complete palm pulled loose, with the sound of separating Velcro, into a giant water blister. I saw pleasant visions of sitting in the shade for a few days while my hand mended. Complaints about such minor injuries fell on deaf ears. So my mirage of a desert oasis popped along with the blister, I put on my gloves and continued to churn.

When we had a series of six or eight holes drilled to a depth of about six feet and about eight or ten feet back from the edge of the overburden, we brought out the dynamite. The first time, I heaved an anxious sigh and whispered to the wind, "Brother Bob, I pray you know what you are doing."

The first step was to spring the holes. That meant that we needed to increase the size of the bottom of each hole so we could cram a dozen or more sticks of dynamite into it. A half stick of dynamite in the bottom of each hole created the breath of "spring" that we needed. Bob would cut a short fuse, insert it into a blasting cap and push it into a half stick of dynamite, light it and slide it down the hole with a muck-stick. In a few seconds, *WHAM*, mud would shoot vertically out of the hole and rain down on us. If we hadn't removed the muck-stick, the blast would send it into low earth-orbit.

Once we had sprung the bottoms of the holes to about the size of a gallon jug, we could then load them with the prescribed number of sticks. Bob started by slashing the cardboard casings of the dynamite, so each stick would compress, like an accordion when you pushed on its ends. We would slide them, one after the other, to the bottom of the hole. Bob then compressed the dynamite by tamping it with a muck-stick with much more force than I would have used. In fact, while Bob was tamping dynamite, an observant javelina might have seen me backpedaling toward the safety of the next mountain range. I sometimes pondered if the world would even know if two grubby young fellows from Nebraska blew themselves to bits in the Arizona desert.

Photo 11: San Pedro River Valley, Arizona—Near Redington Camp
—Jan 1962

Relieved that no loud sounds had yet interrupted the silence, I watched as Bob inserted fuses with detonating caps into the final sticks and pushed them firmly into each hole. We then gathered up enough mud to tamp the holes shut so there would be no escape for the expanding gasses when the blast went off. Bob had cut the fuses to staggered lengths to compensate for the time it took to light each hole. This would allow all blasts to go off at about the same time. The fuse lengths were way too short for my conservative nature. In reality, we didn't need the time that Bob allowed. It is surprising how fast you can move when there is "fire in the hole" right behind you. When all fuses were sizzling away, I could always outrun Bob to the safety of the next ridgeline, where we would wait and wait and wait.

When the wait had become almost unbearable, "*KABOOM*," a dust cloud would fly into the air and we would duck as potato-sized clods flew off into the desert.

Once the breezes had depleted the noxious blasting fumes, we would return to the quarry. Now all of the overburden was broken into chunks that we could manage. Like ants in an anthill, we would shove and roll the large blocks of clay and stack them in a wall behind us in the quarry. We could then backfill behind them with the smaller materials. We usually accomplished the backfill with a shovel—many, many swings of a shovel.

When we had finished removing overburden, we had exposed an area of perhaps thirty feet by eight feet of the layer that contained the fossils. After we processed the layer with small picks and hand-tools and removed the fossils, we would start all over again.

As it is impossible to get the same amount of materials back into the hole from whence it came, our pile of removed material behind us grew to a mountain larger than the one we were digging away. We soon had only a narrow trench to work in, and a ten-foot wall to throw our dirt over. The fun of working out the fossil layers and discovering new things was worth the manual labor, although I am still immensely fonder of the former.

When we moved from quarry to quarry, we carried the extra dynamite with us, behind the seat in the jeep, as if it was as benign as our bedrolls. After a few months of sleeping in our bedrolls without benefit of laundry or bath, that's not saying much. I was surprised that Bob had learned so much about the art of dynamiting during his winters with Ted Galusha. Nonetheless, it was all in a day's work back then. If two young grubby fellows from Ainsworth, Nebraska tried to purchase dynamite in Tucson, Arizona today, they would certainly invite a thorough interrogation from the Tucson City Police if not the FBI.

Bob and I spent our evenings, in front of a blazing mesquite fire, playing cribbage and watching foxes and other night creatures slip up to satisfy their curiosity. When the quarry was too wet to work after rains, we found other things to occupy our time. For something to do, Bob fashioned a sling, like the one with which David killed Goliath. I believe he used the leather tongue of a worn-out boot and a couple of shoestrings. He loaded the boot-tongue with a golf-ball-sized rock and, after a healthy windup, flung it right through the side of our tent. Bob found something else to do.

Later that winter season our father came to spend some time in camp with us. His recent history had included an emotionally draining divorce. Time in the Arizona sun heals many ills. He joined us in the fossil quarries and, for a short time, Bob gave orders to his dad. If Fritz Skinner reads this, I am sure he is now very envious of Bob. Oh, for such an opportunity!

It was in those fossil quarries that both Bob and I quietly decided to go to college. It was clear to me then that Bob was on the unrelenting trail of the fossil. While digging bones out of a hole in the ground, I would witness jet trainers and fighters from various Air Force bases zoom overhead. I decided that upstairs was where I wanted to be. It wasn't because of my lack of enthusiasm for paleontology that I made that choice. We were heading for trouble in Vietnam and the draft was in full operation. I registered for the draft in Prescott, Arizona on my 18th birthday so the military was often on my mind that year. If Uncle Sam decided to offer me up for target practice in Southeast Asia, I wanted speed on my side. To fly for the Air Force, I needed to be a commissioned officer and that required a college degree. Bob and I sent our applications to Colorado State University. Bob's application was routinely accepted—naturally! CSU took a cursory look at my application and told me that they had just about had their fill of out-of-state students. My dad sent a letter off to the Dean to explain that I wasn't nearly as dumb as my high school credentials had indicated. Good old Dad had a way with words. The Dean took another look at the same application and decided that CSU could use my tuition money after all.

Bob transferred to Ted Galusha's field crew during the summers, continued through graduate school at Columbia University, and gained deferments from the draft. He applied for a position at the Smithsonian Institution's National Museum of Natural History and was accepted. After a few more years of simmering, the Vietnam War reached a boiling point. Ten years after my winter in the Sonora Desert with Bob, Uncle Sam indeed offered me up as high-speed skeet in the skies of South and North Vietnam, Laos, and Cambodia.

Marie wrote about Bob in a letter to me. "We often said of him, 'Whoever gets Bob will have a successful fossil-hunting season!'" She stated her impressions even more emphatically in an earlier letter. "Anyone who had Bob working with them would have the best collection." She continued, "He was the first one in his Columbia class to pass his final exam to earn a doctorate. We were very proud of him. Bob's life at the Smithsonian is a mirror of a world of accomplishments." Her recent letter that mentions Bob, states, "We all looked up to Bob and so did everyone we knew. His grasp of geology in the field without formal training was phenomenal, as was his mild unpresuming manner. He will always be my favorite."

There is more about Bob's youth and our brotherhood in the final section, *Bone Digging Days*. Bob was ethical, painstaking, and responsible then. I am sure those qualities have not eroded with forty more years of practice. You may learn more about my impressions of my brother in the pages that follow. If you want to know Bob's impressions of *me* or his recollections of *his* ongoing bone digging days, look him up at the Smithsonian Institution. You may find him there or in the Republic of Georgia, or in Kazakhstan, China, Mongolia, along the blue Danube, or wherever the elusive mammalian fossils hide.

I was in *fine company* when I was in a fossil quarry with Bob in 1961. I was in *fine company* in the outhouse hole with Fritz. I will always consider myself to be in *fine company* whenever my memories bring forth images of those golden days of bone digging and the folks who went to the field.

PART 3

BONE DIGGING DAYS

Photo 12: The author, Raleigh Emry, at age sixteen.

Morris F. Skinner stopped by my parents' door one morning in 1960, and invited me, a sixteen-year-old kid, to be a fossil hunter. Except for my parents, Morris influenced my life, and for the better, more than any other person. He took me under his wing and showed me how science is applied in the field. He taught me the geology, stratigraphy, paleontology that he knew so well. He insisted that I learn these things, as they were important to my job. Nonetheless, he didn't urge me to follow his path through life. He just presented his profession in a compelling way. If my earlier interest in aviation, the Selective Service Draft and the Vietnam War hadn't swayed me to make other decisions about my future, Morris would surely have won me over. He brought meaning to weighty words like dedication, tenacity, and conviction. Moreover, he brought meaning to the words that are as light as air—laughter, joy and friendship. These are some of my fondest memories of my *Bone Digging Days*.

Chapter 15

Photo 13: Niobrara River, Nebraska—Fall 1959

Elephants Just Upriver!

Morris Skinner has been a familiar name to me since I was about eight years old. My first encounter was on a summer day in about 1952. The Emry family was living on a small ranch in the Niobrara River canyons, about seventeen miles and at the end of a trail road northeast of Ainsworth, Nebraska.[27] My parents, two sisters and one brother eked out a living on a rustic little spread that was too little grassland to be a prosperous ranch and too little farmland to be a prosperous farm. I prefer to call it a ranch because Roy Rogers and Gene Autry didn't cotton much to sodbusters. In spite of the name, our spread was an abundance of canyons. The canyons lured Morris Skinner there.

I had been fishing in one of our ponds that my dad had stocked with trout. I walked to the house, very Tom Sawyerish, with my willow pole over one shoulder and carrying a stringer of one trout, perhaps a foot long. (At this early point in my story, I am tempted to write, "three trout about eighteen inches long." However, I am quite certain that it was a stringer of one trout perhaps a foot long—perhaps shorter—but definitely not less than one trout.)

I had not yet realized that we were having the rare experience of visitors. I was nearly as shy as the deer that hid in our canyons. Our ramshackle set of buildings, tucked away in trees and hidden from view, were as secluded as they could possibly be. My family was ardently self-sufficient but unconditionally poor. Only a few years before, the renowned Blizzard of 1949 came with a vengeance and left us snowbound for six weeks; I hardly recognized the inconvenience. From my perspective, the continual snowstorm simply made for wonderful sledding down the steep trail into the canyon. I was a young boy on a ranch where all of the typical domestic animals roamed. In addition, we also had nearly every kind of wild animal pet you can imagine. At any given

---

[27] The ranch of my youth is located at the point where US Highway 183 now enters the Niobrara River canyons from the south. The highway was not yet built when we lived there. Access to our ranch was by section-line roads and then trail-roads through neighboring pastures for the last few miles. The photo at the beginning of this chapter is looking upriver from a hill a short distance downstream from my boyhood home.

time, we had pet raccoons, badgers, bobcats, skunks, packrats, ground squirrels, tree squirrels, magpies, crows, sand-turtles and others. To me, our hardscrabble ranch was about as close to heaven as I could get.

I often rode my Shetland pony to our one-room schoolhouse and Joy Pense Jones, a wonderful lady who had once taught my grandmother, taught me through fifth grade. Only two other students attended South Riverside District 26, during my last year there as the others had gone off to high school in Ainsworth. My youngest sister, Ruth, a neighbor girl, Linda Conrad, and I were the youngest of the "river rats" and the last to attend the old school. Because of our remoteness, there were no more than two dozen faces in the entire world that I could put a name to.

People who came into our canyons were either there for a purpose or hopelessly lost. If I had known we had visitors, I would have continued to fish instead of going through the nervous formality of being introduced to strangers. So being unaware there were outsiders around, I approached the house with my solitary trout.

A water tank near our kitchen door supplied us with pure, cold spring-water from a pipe in the hillside. As I neared, I noticed two men with my dad. They sat in the shade by the water tank and talked as they passed the dipper around. The strangers wore tan clothing and carried packsacks on their backs. They carried small shovels with short "D" handles. They seemed out of place in a land of Levi's and cowboy shirts, or if you cotton to sodbusters, a land of bib overalls.

My fish was a conversation piece. The two strangers were curious where I had caught a trout. The two nearest trout streams, Plum Creek and Pine Creek, emptied into the Niobrara River several miles above and below our ranch. Presuming that game wardens might dress in tan clothing, I hesitated while I mentally reviewed fishing laws. Satisfied that my catch was legal, I held out my catch as Dad explained that it came from our pond. They inspected my trout and commented in a friendly way about its beauty. I became sufficiently curious to stick around to listen to them talk to my dad.

One of the men was short and sturdy. The other was tall and lean. I learned later that the former was Morris Skinner, the latter, I believe, was Ralph Mefferd[28]. They were asking my dad about fossils. They wondered if he had found any fossils in any of our canyons. My dad showed them some pottery shards and arrowheads that he had found in a place above the canyon that we called the "Indian Town." The two men showed interest in Dad's archaeological artifacts but explained that they were looking for much older things. I could not imagine that there could be things in our canyons much older than the relics from the Indian Town.

I learned, many years later, that our creeks cut into geologic strata called the Pierre Shale, an upper Cretaceous marine formation. Above the shale, there was an unconformity—missing strata for many millions of years. On top of the shale, there were pockets of Pleistocene gravel and above that, windblown sand of the sand-hills.[29] The Pierre Shale was far too old and the Pleistocene gravel far too scattered to make our ranch good grazing for the early mammalian fossils they were seeking.

One of Dad's many amusements was to peel back layers of shale in our creek-beds. He could find small nuggets of iron pyrite—fools gold—sandwiched between the layers. He enjoyed watching the wide-eyed reactions of gullible witnesses to his prospecting. He had some of these golden nuggets in a fruit jar that he brought out to show the men. They were clearly not gullible. Dad was therefore not able to tally checks

---

[28] Ralph Mefferd was no longer working for Morris at the time of the visit to our canyons. Ralph may have just been out with Morris for the day or the fellow with Morris may have been someone else.
[29] There may be other formations present. However, there are no major outcrops containing mammalian fossils.

in the "fooled" column of his mental score-sheet. The two men correctly identified the mineral and politely took some as samples.

They explained that they had been prospecting farther west and had driven out our way to study the geology. They had parked their pickup somewhere above the canyons and had arrived at our spring in need of a drink of water.

Up river, to the southwest, two and one-half miles as the crow flies, was Devil's Gulch, a well-known fossil locality. My parents had told me of scientists who had collected fossils there and had shipped them to museums all over—Denver, Chicago, even New York City. These men identified themselves as two of the scientists who had worked in Devil's Gulch. In fact, they had been in Devil's Gulch earlier that day.

Devil's Gulch is very near the trail that took us to town. I was quite familiar with the name. Nearly every time we passed by, someone would say, "Well, there's Devil's Gulch." It didn't require a verbal response. Nonetheless, I always looked to see if the Devil was in. A windowless, old house teetered near the edge of the gulch. It seemed to be a likely place for Lucifer to hang out. I could imagine that he gobbled up critters and threw the bones into the ravine that bore his name. Scientists were the brave souls who sneaked down there to haul the bones back out.

The short, sturdy scientist opened his pack and took out some interesting things to show us. He explained that some dark objects about two inches long and somewhat square on the end, were teeth from an extinct horse, possibly a zebra. He showed us the patterns in the grinding surfaces of the teeth and explained that they were unique to each type of horse. Experts could sometimes identify the type of horse by those patterns alone. He then unwrapped some irregular chunks of white curved fossil that reminded me of pieces of a broken pickle crock. He explained that they were bits of turtle shell from turtles as big as those that live on the Galapagos Islands. He then showed us an odd shaped specimen and explained it was a toe bone of an extinct elephant called a mastodon.

The toe bone reminded my dad of a doorstop that he had brought home from his own hike into Devils Gulch. It was a hemisphere about eight inches in diameter. Morris quickly identified it as the ball from the end of a mastodon femur—the ball that fits into the socket of the pelvis. As it was an isolated piece, it had little scientific value, so we kept it as a doorstop.

I had heard many unreliable stories. In fact, Dad had guessed correctly that our doorstop had once belonged to an elephant. This was the first time a scientist had confirmed that there had been elephants just upriver!

Chapter 16

Brother Bob Breaks Trail

In 1957 and about five years after my first encounter with Morris Skinner, my brother Bob graduated from Ainsworth High School. As more years passed, he became known in scientific circles as Robert J. Emry, Ph.D., Curator of Vertebrate Paleontology, National Museum of Natural History, Smithsonian Institution, Washington, D.C.

During Bob's childhood years, on our ranch in the canyons, he had suffered from asthma. Sometimes our parents could not disguise their fears that his future may be short. There were many nights when Dad or Mom would take Bob out of the muggy canyons and to a high hill in the pasture where he could breathe more easily in the prairie breeze. Bob slowly overcame the severe asthma attacks that rendered him weak and thin. Summer in Nebraska is unfortunately a terrible time for someone who is predisposed to allergies.

What Bob lacked in brawn he made up for in brains. One of his educators, along the way, tested Bob's IQ. His sky-high score forecast a bright future for him. Bob's breaking the curve, unfortunately, brought me several bouts of high-school misery. Filling his shoes was easy when they were hand-me-downs—if he had left enough leather to tie onto my feet. Filling his academic shoes was another story. When Bob didn't go to college immediately upon high-school graduation[30], his teachers were dismayed. For the four years of my high school (I graduated in 1961), they regularly inquired about Bob, "Has Bob gone to college yet?—No?—What a waste of genius!"

Conversely, I showed a modest interest in higher education during my senior year and it went unnoticed. Seeking encouragement, I asked my favorite teacher how I could fulfill my dreams of becoming an airline pilot. I hoped I would hear some of the following, "go to flight school—get an education—go to college—you can do it!—don't waste *your* genius!" I knew that some of those words were in my teacher's vocabulary as I had heard them used on several occasions in reference to my brother. Instead, his recommendation is a memory that hasn't faded or evolved, "Why not go down to Omaha and get a job sorting luggage at the airport?"

From my sour-grapes point of view, I was convinced that the teachers had all fallen for the Pygmalion Effect. I presumed, because of Bob's measured high IQ, they expected brilliance and Bob's schoolwork often fulfilled their expectations. His occasional less-than-brilliant efforts were easily overlooked.

It seemed that the antithesis always happened to me. Teachers expected mediocrity from me, and my schoolwork often fulfilled their expectations. My occasional flashes of brilliance were easily overlooked.

When I returned for my twentieth reunion, my history teacher looked me up and down with a curious eye and finally recognized me, "Of course! You are Bob's brother!"

The summer after Bob graduated from high school, he continued his high-school job of pumping gas at the local Conoco and trying his best to cope with the allergy season. Our dad knew that Morris Skinner hired local boys as field assistants on his bone digging expeditions. Our

---

[30] Bob and I both enrolled at Colorado State University, Fort Collins, Colorado, in the fall of 1962.

parents hoped that if their oldest son could get away from his allergies, his health would improve. Dad asked Morris Skinner about hiring Bob and the rest is history.

After Bob's first summer in Wyoming and winter in Arizona, he returned home handsome, tanned, fit and muscular—just as my parents had hoped. I then had to deal with an older brother who was not only a genius, but he was handsome, tanned, fit and muscular to boot!

As I am the youngest of my siblings, it seemed to me that in those early years I was always in Bob's shadow or footsteps. Although I sometimes resented it then, I didn't realize he was simply leading the way. I am very proud of Brother Bob. I appreciate his breaking trail for me.[31]

---

[31] Bob also saved me from drowning in the Niobrara River when I was a lad. I had been swept under a tangle of logs and brush while swimming. I caught the upstream log and held on for dear life as the current pulled my body under the trash. The water was breaking around my head when Bob pulled me out. I will never know if I would have lodged in the brush and drowned or if I would have safely passed to the other side. I am forever grateful to Bob for keeping me from learning which fate was in store.

Chapter 17

Pi Times Odometer

It was in 1960, just after my junior year in high school, that I became a field assistant to Morris Skinner. My employment, up to that point, had been as a janitor's assistant at the high school (not an easy way to impress girls), as a summer laborer in a hayfield (too much like being a sodbuster), and working for my dad on the ranch (too much like work). I had yet to line up suitable employment for my last summer of high school.

Morris Skinner had a predicament. Due to life's little circumstances, he was short one field assistant. I am not sure why he selected me to fill the vacancy. I strongly believe that my sterling reputation preceded me (that, of course, is my memory of record). It may have been a covert effort by my parents to get me the heck out of the kitchen. I didn't suffer from asthma, but I had a serious ailment called *insatiable appetite*. My voracious phase might have been a good enough reason for Dad to play matchmaker again. Morris, being pleased with Bob's employment, may have thought there was a good chance that the same tree could produce two good apples. So, one of those reasons or none of those reasons brought Morris to our door.

We lived in Ainsworth then and Dad commuted to the ranch each day. City living was at first a temporary arrangement while we kids were in high school. As time went by, it became permanent.

We rented a tiny basement apartment on North Pine Street. The builder, years before, had started a house and quit after he had laid the basement foundation. He capped it off with a flat, tarpaper roof, about three feet above ground level, and called it complete. The highest part of our dwelling was the entrance at the top of the stairs that extended high enough to frame a door. I suppose the exterior dimensions were about thirty feet long and twenty-four feet wide. Within that tiny space were two bedrooms separated by a closet, a bathroom, a utility room, a pantry under the stairway, and a combination kitchen/living room.

Not only did my family of six live there, but also two cousins and a few friends were semi-permanent residents. Thankfully, our "badger den home" was reasonably warm in the winter and fairly cool in the summer. It also had several luxuries that we never had in our old house on the river. Propane gas now cooked our meals and heated our water. An oil-burning heating stove kept the place cozy in winter. Therefore, we no longer needed to cut or haul a single stick of firewood. Indoor plumbing meant we could also stay inside for all of life's necessities. A hot steamy shower was so luxurious and new to me that I didn't consider it a bath; a cleaning chore could never feel that wonderful! Moreover, what could be better than cold milk direct from the refrigerator without first milking a cow? High on my list of new and wonderful novelties was a simple ice cube. With such luxuries, it was easy to see why Dad could not persuade some of the Emry family to go back to the ranch.

When I answered the knock on the door of our luxury suite, Morris offered me six bucks per day to join his crew.[32] Although I could vaguely picture the magnitude of six dollars, I had no idea what "per day" meant. I soon learned. For the spoiled youth of the new millennium, I will

---

[32] At the start of my seventh year, Morris doubled my starting salary to twelve dollars per day. Of course, I was a college graduate then. I suppose that contributed to my inflated wages.

explain it. "Per day" began at the slightest inkling of light on the eastern horizon or when my air mattress went flat—courtesy of Morris. "Per day" ended when we had squeezed the last photon out of the sunset—plus a little while longer. Per days were l-o-n-g in those latitudes.

Morris' offer of employment was contingent on my being ready to leave in an hour. I could easily fit all of my belongings into an old Army-surplus ammunition can, so I met his conditions with about fifty-eight minutes to spare.

My mom offered a mild protest that he was robbing her of her baby. She had worked long hours at Spearman's IGA, a little corner grocery on Main Street and later as a telephone operator. Her thirty-five cents-per-hour was stretched thin. Mom may have calculated the number of hours of such busy work it took just to keep *my* belly full. She also knew that my big brother had thrived under Morris' supervision. Consequently, her protest was not convincing. She asked Morris where we would be going.

Morris answered, "Up to the Pine Ridge and then on to North Dakota." Pine Ridge is a physical landmark as well as an Indian Reservation. Our government gave the Native Americans some land for which they could not contemplate a better use. It is similar to my parents' ranch—more vertical than horizontal. Early explorers called these features "badlands" and they extend not only through much of the Indian reservations, but throughout the western Dakotas, northwestern Nebraska, and eastern Wyoming.

My mom looked tentative while she contemplated Morris' answer.

To assure my mother that her tender son would be safe under his guardianship, Morris waved his hand as if to cancel any doubt. Mom looked relieved, but only for a moment. Morris had a saying for every occasion. If a tried-and-true adage was not on the tip of his tongue, he would, and sometimes with insufficient thought, make up one on the spot. In bravado exaggeration, as if he were playing a buckskinned scout leading an exploration into the uncharted frontier fraught with hostile natives, he drawled something that was certainly amusing but several shades off-color. It had to do with protecting my virginity.

Morris had just committed his first "Skinnerism" in my presence. I felt a blush conquer my face, but I didn't know whether to guffaw and please my potential employer or look shocked in deference to my mother's sensibilities. Miraculously, the whole thing blew over in a few seconds as if it had been a surprising little whirlwind, the kind that lifts your hat from your head and spins it around. The Skinnerism spun out as quickly as it was spoken.

Mom's slight smile of amusement sealed my employment. She gave me a hug, told me to be good, and warned me to watch out for the many things that mother's warn their sons about. Brother Bob had already whetted my appetite with tales of his adventures with Morris. That moment, in our driveway, was one of the happiest of my life. I knew I was in for an interesting summer.

When Morris, Bob and I climbed into the pickup, a 1957 "government-green" Ford, I glanced at the odometer. I had never been very far from home so I wanted a reference from which to gauge my summer travels. One of the few things that I had learned in Geometry, Algebra and Trigonometry was that Pi equals 3.1416. The digits on the odometer read 31416. This was only a coincidence, of course. However, if the odometer had read anything else I would not have remembered it as a milepost in my life—the day I first left home to become a bone digger.

Chapter 18

Sir? You'd better come out here!

Meaningless numbers quickly replaced pi on the odometer as we drove to Morris Skinner's headquarters on the south side of Ainsworth. We had to pack the pickup before we set out for the Pine Ridge and then North Dakota, as he promised.

The packing shed was a converted barn and pig shed. It held all sorts of interesting things. Deer and elk antlers and buffalo skulls decorated the exterior. Sometime earlier, during the barn's conversion to a packing shed, Morris had poured a concrete floor. Several things had been memorialized in the wet concrete. I especially remember "Fritz, May 6, 1933" as I was born on Fritz Skinner's birthday some eleven years later. Fritz is the nickname of Morris F. Jr. Other names, dates and handprints were there too. Near the back wall, there were two, oval impressions in the concrete—overlapping indentions about the size of watermelons. This impressionistic artwork, signed by its creator, M.F. Skinner, is clear and eternal evidence that Morris' Indian name "Hosay Tonka—He of the Big Butt" was appropriate.

Before we loaded the pickup, Morris looked at me with a devilish leer and asked, "Have you had your Rocky Mountain Spotted Fever shot?"

We were several hundred miles from the Rocky Mountains, and I had never been much farther than the county line, so I was sure I hadn't. Even though I could count all of my previous inoculations on two fingers ("two bared buttocks" would be more anatomically accurate), my instincts told me that something ending in "shot" had something to do with a needle. I warily shook my head.

I supposed Morris was about to take me down to Dr. Shiffermiller's office. Instead, he took me to the kitchen. I discovered that Morris always insisted on administering our inoculations, even when his medical doctor son, Fritz, was present. Morris opened a tin box full of archaic paraphernalia, and began to screw it all together. When he had finished, he had constructed a sizable syringe with an impressive needle. He proceeded to boil it in a pan to sterilize it, retrieved a serum bottle from the refrigerator, and with little fanfare protected me from the demon tick. His technique was to jab the needle deep into upper arm muscle and squirt the serum in with a swift push of the plunger. "Deep into the upper arm muscle" for me was only a half-inch or so before bone or other tissue interfered. Although such obstacles didn't seem to concern Morris, the resulting aching welt concerned me for a few days whenever anything touched my arm with more force than a light breeze. Nevertheless, I looked on the bright side. A Rocky Mountain spotted fever inoculation just might foretell my seeing the Rocky Mountains that summer!

Packing the pickup was a standard routine. Behind the front seat, we placed cardboard folders of topographic maps. Morris had modified the step-wells at each door to hold small ammunition cans. The Army-surplus cans were watertight and ideal for camera equipment, or tools. Morris had even put hinges on the interior door panels that hid the workings of the window cranks and door latches. This was a route to extra space in the bottom of the doors to put seldom-used essentials.

One ammunition can held a spark-plug tire pump. By removing a spark plug from the engine, inserting the pump into the spark-plug hole, and screwing an air-hose to it, we were in business. Once we cranked up the V-8, we could inflate air mattresses and tires quite easily. I was concerned that the mixture in my air mattress might be volatile and I would blow sky-high some night. A tag on the little pump indicated it pumped fresh air and not a fuel rich mixture from the cylinder, but I sometimes doubted what I had read.

Morris stuffed a plastic cigarette case in a nook behind the overhead trim. Inside was a roll of money sufficient to pay for the expenses of our trip.

We put spare tires in first and held them in place with the tent box, a long wooden box with a metal covered lid. Then we loaded the "big box." The big box was as long as the width of the pickup bed and about three feet wide and two feet high. It held rubber bags of Plaster of Paris and an ample supply of burlap sacks, newspapers and other supplies. We used those items for casting fossils. We also threw our heavy parkas into the big box. We never knew when the weather would turn cold. We buried frozen chunks of meat into the mass of big box items. If we carefully wrapped the meat in shredded newspaper and buried it in the big box, it would not thaw for several days. If we forgot it there, it would be sure to remind us within a week. We placed picks and shovels on both sides of the boxes and in the spaces above the wheel wells.

We often placed our bedrolls on top of the two boxes. They made decent seat cushions for those of us who rode in the back. Our sleeping bags were our personal responsibilities. Morris supplied an outer cover of canvas for each of us, as we seldom used the tent. Instead, we rolled out our bedrolls on a tarpaulin and slept under the stars. If mosquitoes were pesky, we put up mosquito nets.

I borrowed a bedroll from Morris for my first trip and sent an order to a mail-order catalog for a sleeping bag. "Cheap" was my only requirement. When it came, I was pleased that its tag stated, "Tested on Mt. Whitney." After the first chilly night when I nearly froze in my fart-sack (a charming name for our sleeping bags), Morris ribbed me about my fine purchase and its claim, "Tested on Mt. Whitney, huh?"

I defended my fine, new possession, "I'm sure it was tested on Mt. Whitney. The tag just doesn't say how warm the guy slept!"

We threw our backpacks on top of the bedrolls. If not, they went wherever we deemed convenient. Our prospecting packs contained a shellac bottle, a knife and file for cutting the burlap and trimming the casts, newspaper and manila tape for wrapping smaller specimens, whisk brooms, hooks and awls for preparing fossils for casting and collecting. We also carried a roll of toilet paper, a snakebite kit and other miscellaneous items.

Morris had the packs handmade to his specifications. The main compartment of the pack could open wide and accept large, irregular objects to accommodate carrying fossil casts of odd dimensions. The packs did not have frames or padding. Only a thin layer of canvas separated our backs from the lumpy masses we carried. On long climbs out of badland ravines, the heavy, hard casts would call out cadence by rocking back and forth across our spines with each step. We tolerated this torture, as that is just how it was done. We needed our energy to power our legs so there was no point in wasting any by complaining. In fact, we took great pride in being tough enough to take it. I was often humiliated when Morris would appraise my spindly legs and select the biggest casts for himself. Nevertheless, I suffered in silence and tried not to let my wobbly legs betray me when I finally reached the pickup with my inferior load.

We seldom carried drinking water in our packs. Morris would insist that we all drink water like camels, before we ventured out to prospect. His axiom was, "No sense carrying on your back what you can carry in your belly." As Morris' girth was twice mine, I sometimes thought, "Well, easy for you to say!" By lunchtime, or quitting time in the evening, we were always more than ready for a drink.

In my later years of employment, I became daring and defied the axiom. I covered an empty, quart whiskey bottle with burlap and used it for my canteen. On rare occasions, when the heat was intense and the pickup was far away, Morris would ask, "Hey, Raleigh?" I would answer. He would continue, "Got any water in your jug?" When I would slosh it as evidence he would ask, "Can I have a swig?—No sense carrying on your back what I can carry in my belly." The axiom evolved slightly.

We put our personal clothing containers wherever there was room in the pickup. Most of us used a large ammunition can. They were waterproof and could be set out in the elements. My camera and film went in there as well.

We placed the grub-box toward the back of the pickup. It was a large box, but smaller than the big box. We filled it with all of the non-perishable food that we would require on the trip. We packed plenty of "bully beef" and "average beans." Bully beef was the term we used for canned, corned beef. Morris deemed a bean "average" if it produced four farts.

I won't dwell on this F-word of a minor chord, but it deserves explanation. When in refined company, the art of passing gas should probably be called "F Flat" or "F Sharp" instead. Morris devised a scientific classification of these passing wind conditions. If one drifted by on the breeze, Morris could accurately put it into one of the following categories: Fizz, Fuzz, Fizzer, Poop-a-nanny, Poop, Tear Ass, or Rattler.

Along either side of the pickup box, we had two folding camp tables, the type with four seats that fold into them. We anchored them in place with a ten-gallon cream can of water.

We put water jugs in bins between the rear wheel wells and the tailgate. We would procure empty, one-gallon glass jugs from the local cafes. They once held syrup for soft drinks. We then sewed a burlap bag to the outside of each jug. A beginner's jug looked like a deformity. After a few years of practice, the jugs looked like a seamstress had fashioned them. Their appearance was not important as they rarely lasted more than a few trips before they became casualties of taking big bumps too fast. We usually left town with six jugs of water. When they became empty, we filled them from the cream can or at the nearest windmill or service station.

The seldom-abused axiom of bone digging life was "Water is only for drinking and casting fossils and not necessarily in that order." We drank from the glass water jugs. We often carried water for casting fossils in five-gallon, Army-surplus, rubber bladders. After the water stagnated in the sun in those rubber bladders for a few hours, it was not suitable for much else. If you tried to drink it, the caustic bitter rubber would destroy your tasting abilities for quite some time—except for the enhanced and lingering taste of caustic, bitter rubber.

We placed two Coleman camp stoves in the top bins of the sideboards. Behind them, we placed several gallon cans of stove fuel and shellac. We also placed other boxes of supplies up there. As we used up the supplies in the upper bins, we filled them with fossils that we had found.

The sideboards had external compartments on each side of the pickup. A latch secured a door as long as the sideboard. When we swung it downward, the door provided a narrow surface to set things on. Morris designed the compartments behind the doors to hold cigar boxes in stacks of four by three. The cigar boxes were ideal catch-alls. Sometimes we used them for small fossil specimens. Sometimes we used them for our own personal items. Our dishes, coffeepots, and other camp necessities were also stored inside these drop-down doors.

I believe I now have the pickup loaded about as we did back then. What you may have noticed by their glaring absence is:

1. Ice: There was no point in taking something that would change into something else before we got ten miles out of town.

2. Insect Repellant: The only varieties that worked back then were oily and unpleasant. After a week, or two, of religiously obeying the axiom about water usage, we really couldn't afford to wear any more layers of oily and unpleasant substances than the layers we had naturally collected. I can remember times when rainstorms chased us into the shelter of the pickup cab. After the three, or four, of us had traded our pungent, steamy aromas for a few minutes, we would sometimes fling open the doors, bail out like paratroopers, and accept the shower for what it was. The positive side of our bone digging B.O. was short lines

in the grocery store. When we went to town for supplies, other shoppers allowed us plenty of space and priority. In those days, many grocery stores gave out Green Stamps, Gold Bond Stamps or other savings coupons with each purchase. If you saved enough stamps, you could trade them for items in a catalog. Morris never bothered with this windfall of savings. Therefore, when the clerk offered the stamps, Morris would have one of us seek out the most needy-looking customer and present the gift. The lucky shopper, not knowing our intention, was sometimes motivated to retreat as we approached.

3. Plastic water jugs: There was no such thing then. Years after my bone digging days, people discovered they could buy, for a heady price, the same stuff that ran freely from their water-taps. This status symbol required a new packaging concept, plastic water jugs.

4. Plastic shellac bottles: A squeeze bottle, the type mustard comes in now, would make an ideal shellac bottle. We used half-pint whiskey bottles for shellac. The metal bottle caps often became shellacked so tightly to the glass necks of the bottles, that we risked ruining our teeth removing them. Morris didn't allow drinking alcohol in camp. Therefore, by the end of each expedition we would need a new supply of half-pint whiskey bottles. Morris was often the only one in the field-crew of legal drinking age. Hence, that terrible burden of procuring shellac bottles was sometimes his and his alone whenever we went back to civilization. Perhaps that is where the term "getting shellacked" came from.

Life in a bone digging camp is pure hell without morning coffee. Morning coffee from a tin cup is not pure hell, but you can feel it from there. The tin absorbs the coffee's heat and transports it to your lips where it dangles for the duration of the trip in the form of ugly blisters. Life in a bone digging camp is pure heaven when you drink your morning coffee from a cheap china mug.

The handles on cheap china mugs, however, presented a problem. The loops stuck out at odd angles when we tried to pack them into the sideboard cubbyholes. Each handle took up almost as much space as the mug itself. My description of loading the pickup should have convinced you that space was at a premium.

Morris discovered a simple solution. By using his pocketknife as a mallet, he could give the handles a quick rap and they would separate cleanly from the mug. Life in a bone digging camp is *almost* heaven when you can drink your morning coffee from a handleless, cheap china mug. Almost heaven? Close enough.

We visited the Ainsworth five-and-dime where they had an ample supply of cheap china mugs. Morris brashly asked the young clerk, a high-school girl I knew, "Are these mugs good ones?"

She helpfully nodded, "I believe so."

Morris said, "Let's see." He then picked up a mug and gave its handle a rap with his pocketknife. The handle clinked away and fell to the floor. Morris exclaimed, "This'n isn't" and set it aside. He picked up another and rapped it with the same result and with the same exclamation, "This'n isn't." He picked up another…

The girl, whom I knew, but not as well as I once had hoped, looked utterly stunned. I was quite pleased at her reaction. I believed she was not necessarily prejudiced toward my previous janitorial profession. Instead, I thought that I fell short of her standards for several other obscure reasons. Therefore, I enjoyed her momentary misery at the hands of a master.

She had no way of knowing that the mugs that Morris "broke" were the very mugs he was going to buy. She quickly disappeared toward the back of the store, and in a worried voice, quietly urged the manager, "Sir? You'd better come out here!

Chapter 19

Photo 14: Reno Ranch, Wyoming—pickup in badlands—July 1962

What that pickup needed was a logo!

You may recall that the pickup was "government green." Morris had painted his homemade sideboards the same color. The color had nothing to do with our employer. The Frick Laboratory and the American Museum of Natural History were not government agencies. I suppose a "government green" pickup was in stock when Morris went shopping and he bought it. Nonetheless, I felt the color gave it a perceived official status. I had a brand new driver's license and a strong yearning to use it. I recognized that I was at the bottom of the pecking order when it came to drivers but I remained ready to jump behind the wheel.

The pickup had a big V-8 engine and a manual transmission. We informally called the lowest gear the "granny-gear." We didn't use granny-gear on the road. However, when we were off-road, we could put it down in "granny" and grind our way through the sagebrush and rough country. The pickup didn't have four-wheel-drive. Morris' convoluted philosophy was that we could get stuck plenty good enough with two-wheel drive! The logic seemed irrational but there was some merit to it. With just the two wheels to provide power, we were seldom the victims of our overconfidence. Even with the limitations of only two-wheels pushing, we put that old pickup in some very impressive places.

The constant bump and rattle of off-road driving had tortured the springs in the seat until they had finally given up their springiness. Morris was the usual driver, and he outweighed the rest of us. The lack of cushion was therefore most noticeable where the driver sat. The beat-out driver's seat bottomed under his weight. Morris was also shorter than most of us. The combination of the seat and his height severely restricted his visibility. He could not see anything that was at a short distance, directly in front, and, due to the sideboards, he could not see anything at a short distance directly behind.

Posting one of us up in the back of the pickup to watch for obstacles solved this handicap. We beat out a code on the roof of the cab. Morris would say, "If you want me to stop, just whang-bang on the cab." So, like captains on the bridge, we would telegraph our orders to the engine room by, in Morris' words, "whang-banging" on the cab. Golly, it got funny at times!

Photo 15: Cedar Butte, South Dakota—July 1961

In July of 1961, we navigated up a dry wash into a box canyon in the shadow of Cedar Butte, South Dakota where we could go no further. I had been the lookout who had gotten us that far. We stared at the vertical wall in front of us that was clearly unconquerable. Therefore, we slowly maneuvered around until we had the pickup pointing back down the wash. One more ten-foot seesaw and we would be on our way. I thought, by then, that it was universal knowledge that the giant vertical wall was now about twenty feet behind. I saw no need to telegraph further orders. With tires howling and the motor racing, Morris backed, not only the needed ten feet but also an additional ten feet, directly into the great clay wall. A minor avalanche of clay about half-filled the pickup box and a dust cloud rolled down the canyon. I was somewhere in the mix. Morris called out, "Damn it Raleigh! Why didn't you whang-bang?"

On another adventure, Morris and I came upon a car, an ancient Chevy, stalled in the center of the highway twenty miles west of Mission, South Dakota on the Rosebud Indian Reservation. A form was slumped over the wheel. Morris stopped and asked me to investigate.

I reluctantly approached the car and rapped the window with my knuckles. An old man sleepily peeked out from under a giant cowboy hat. He was an elderly resident of the reservation. I asked, "What's the problem?" The blank look on his face seemed to convey there was no problem. In a moment, he pointed to the gas gauge. Without a word from the old man, Morris decided to help. He said he would steer the car and I would push them into Mission with the pickup. The old man tentatively scooted over and Morris got behind the wheel. I eased up behind and off we went. When we headed down an incline I would slack off, they would roll along, and I would meet them bumper-to-bumper on the next upgrade. Sometimes we rejoined smoothly, sometimes with a bit of a jounce.

Morris sat with both hands on the wheel, keeping the old car between the ditches. When it seemed, to me, that our existing speed was working well, I would add another mph or two. There became a finite speed when some internal vibration violently shook the old car. All I saw was a subtle bumping of their hat-brims. I mistook Morris' first quick signal as a sign that we could go a little faster so I increased speed again. Morris finally, and I suppose bravely, loosened a hand from the wheel long enough to convey with great certainty that it was time for me to slow down. After a thorough, one way discussion of the situation, we limped on at a more modest pace. We left the old man sitting as we found him, but in Mission, South Dakota. I was never certain that Mission was where he wanted to go. Except for his pointing at the gas gauge, he had remained stone-faced throughout the trip. He may have tried to tell us he was only waiting for someone to return with some gas. As we drove away, the far-away look in the old man's eye seemed to convey his thoughts that, once more, he had been maltreated at the hands of the white man.[33]

At an earlier time, on the reservation, we were the ones without gas.[34] Whenever we filled the tank, Morris would rock the pickup chassis with one hand as he pumped gas with the other to burp every last air bubble and make room for more fuel. We filled the tank right to the top of the spout. Morris was always calculating the remaining gallons of fuel in the tank depending upon the gas gauge reading. His number crunching revealed that we had gas in the tank when the rest of us presumed we were driving on fumes. We were on our way back from Arrow Wound Table and were several miles out of Oglala, South Dakota when the pickup proved him wrong. The engine started to sputter and Morris rocked the wheel left and right to slosh the remaining gas into the feed line. We finally coasted to a stop.

Always resourceful, Morris got the remaining stove fuel out of the sideboard and poured it into the gas tank. The octane of our stove fuel was somewhere outside the operating parameters of the V-8 engine. After a few minutes of cranking, we were now out of fuel and we had also depleted the battery. Morris ordered us to push and he would pop the clutch to start it. So we would get the pickup rolling, Morris would pop the clutch and the pickup would sputter and die. We reached a downgrade where the ditch sloped away at a greater rate than the road-way, so Morris turned the pickup into the ditch to gain more coasting velocity. He popped the clutch again and the engine finally exploded to life. Its valves rattled and tapped as Morris guided it back up onto the

---

[33] My diary describes this adventure on October 6, 1961. In all honesty, the old man did appear to need assistance. It just seemed that humor always intruded into even the most serious of our intentions.

[34] As recorded in my diary on 16 Aug 1961.

road. We all ran and jumped in. In about a hundred yards, the engine gasped, backfired like a howitzer, and died.

We got out and sat in the shade of the pickup while Morris paced and fumed. The evening was approaching so the shade from the pickup now reached about halfway across the road. We lounged in the shade and waited. Although there was no traffic in sight, Morris scolded us; "Dammit guys, get off the road! A car could run right over you!" We looked again, the several miles both ways. Nothing more than a whirlwind was stirring. Nonetheless, we pulled in our feet an inch or two.

After a long wait, an old car came down the road dragging a cloud of dust. Inside were several adults and about a half dozen kids. Morris flagged it down and, speaking the few Lakota Sioux phrases he knew, bummed a ride. The already crowded travelers good-naturedly found room for one more. Morris squeezed in with his gas can, and soon returned with some fuel. This adventure didn't cure Morris of driving while the gauge sat on "empty," but he never again took it to that extreme.

While I am on the subject of transportation, I should tell you of Morris' experience with the two-wheeled variety. We crossed paths frequently with Reid McDonald, a paleontologist from the University of Idaho. One summer, Reid headquartered his fossil hunting party at a rural schoolhouse near Porcupine, South Dakota. We stopped by Reid's encampment to be neighborly and his crew showed us their two modern contrivances called Cushman Tote-Goats. A Tote-Goat was similar to the old Cushman, balloon-tired, motor scooter and was, I suppose, the first two-wheeled, all terrain vehicle.

While Morris and Reid parleyed, a few assistants went prospecting. They let me drive a Tote-Goat and, although I wasn't convinced of its practicality, I had a lot of fun. It had only one gear and with the twist-throttle full open, it may have reached a top speed of 20 mph.

When we returned to the schoolhouse, Morris decided to take a Tote-Goat for a spin. He got the feel of it as he putt-putted off up the road for about a quarter of a mile. He then turned around, and headed back to the schoolhouse. Once he had twisted the throttle to the peg and had reached terminal velocity, he somehow forgot how to untwist it. He later said that he was "untwisting" his left hand instead of his right. As Morris was left-handed and dyslexic, that makes sense.

It was early in the season and he had just bought a new Stetson. It sat proudly on his head and the brim tipped up in front, due to the wind, like a Gabby Hayes trademark. Somewhere in the quarter mile, he realized he could not negotiate the 90-degree turn into the driveway of the schoolhouse. So, at full tilt he headed on an angular cutoff through the ditch, which was overgrown with wild sunflowers and other weeds. Morris never ever looked panicked or frightened, no matter what dilemma faced him, but there was a distinct terseness on his face as he disappeared into the weed patch.

A washout in the ditch solved his problem for him. The Tote-Goat stopped and Morris continued at about twenty miles per hour. He slid into view with his head tucked under his body. He lay motionless for several long seconds. We observers were a bit worried as we hurried over. Finally, Morris moved and slowly pulled his head out of his shell like a turtle. He had a stunned look. His hat, now crimped into a new style, was covered with grass-stains. As soon as I could see that he had survived, I couldn't control myself. I laughed until I was nearly sick. I told him later that it was such a funny exhibition that I probably would have laughed had it broken his neck. Morris took it all in stride and laughed along—at least as much as his sprained neck would allow.

Whenever we were in Ainsworth that first year, I hoped Morris would ask me to take the "government green" pickup out on errands. Any excuse that put me behind the wheel was fine with me. Once rolling, I could manufacture a route that would take me down Main Street and then in front of the drive-in-snack-bar where high-school kids hung out. I hoped, when I drove by, they would be saying, "Wow! Look at Raleigh! Have you heard he's a field assistant for the American

Museum of Natural History in New York City? He's got a great job with Morris Skinner!" I could imagine all the girls were swooning, "Wow, what a catch!" I knew that my summer job held status-appeal for Ainsworth, Nebraska. However, I was also a pessimist. I could also imagine that no one recognized the pickup as a prestigious museum vehicle. I worried that my schoolmates might think that I still worked as a janitor's helper. I presumed the guys might be saying, "Looks like the school got a new janitor truck. There goes ol' Raleigh again." I was even more worried that the girls might be drinking their cherry-phosphates and sniffing, "Janitor—what a low-life!"

What I thought the pickup really needed was a big sign on each door that stated, "American Museum of Natural History." I could envision the bold print attractively encircling a prestigious logo. That would have been much more valuable to my purposes, that first year, than four-wheel-drive. What that pickup needed was a logo!

Chapter 20

Time to get the old squawk box!

When we were convinced that we had all of the essentials for a bone digging expedition stuffed into every space in the pickup, we would point it down the tree-lined driveway and roar and rattle off into our next adventure. There was one thing very dear to Morris that we usually left behind, his loving wife Marie.

As Marie's full name is Shirley Marie, Morris often fondly called her "Shirley Girlie" with a twist that sounded like a Bronxian "Shoyley Goyley" to me. Their usual friendly banter was always sweet music to my ears. As we roared and rattled down the driveway, Morris would often accompany a "beep-beepity-beep-beep" on the horn with, "So long Shoyley Goyley. See you in a couple weeks!" Marie would stand at the gate and wave us off with a smile and words we usually couldn't hear over the noisy V-8.

Marie did accompany us on our expeditions once or twice each summer. Those were nice times. The days were usually a little shorter, and we ate a whole lot better whenever Marie was in camp. Sometimes we would take short side-trips to interesting places that had nothing to do with fossil collecting. Marie sometimes brought her art supplies along and she would sketch in charcoal or paint in watercolors while the rest of us prospected for fossils. During lunch breaks, we would sometimes

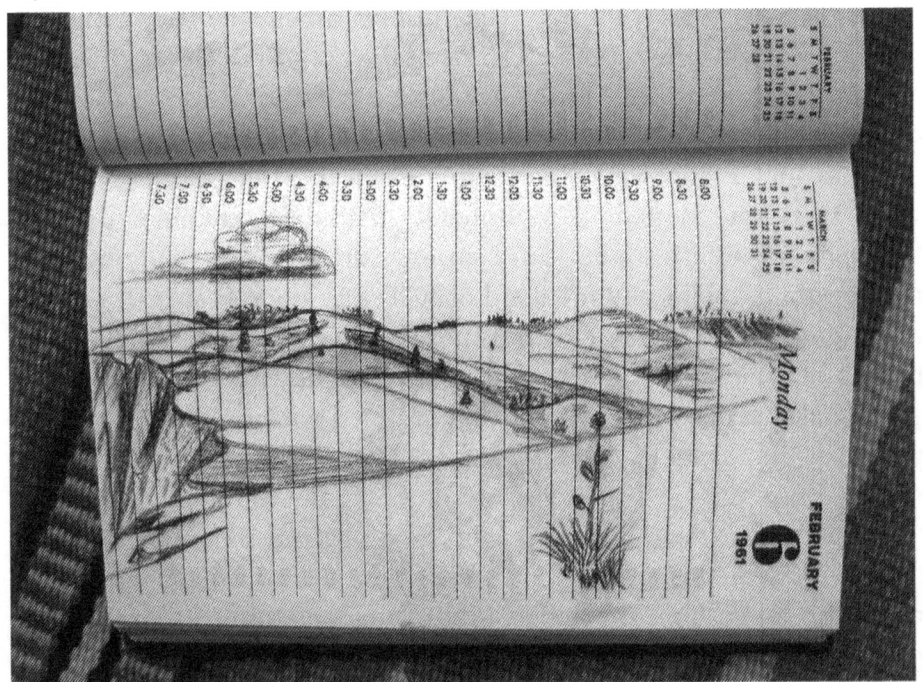

Photo 16: Charcoal by Author—Badland Scene—Sep 6, 1961

wait-out the intense heat of the day in the shade of the pickup. Marie would show me a technique or two and I would scribble charcoal sketches in the blank pages of my diary. My memory remembers my work as beautiful badland landscapes with delicate shading that exactly captured the play of the sun on the land. The evidence, in my diary, looks as if pieces of charcoal had been accidentally lost between the pages and they had shifted randomly with a summer's worth of bumpy roads. I prefer my memory.

When Marie came to camp, she occasionally brought friends along. In the summer of 1962, Francesca LaMont, a lady I thought to be of French heritage, accompanied us. Marie now tells me that Francesca descended from Russian nobility. Francesca lived in New York and was an ichthyologist by training. Even though fishes were her expertise, Francesca was curious about all facets of the natural world. Elva Rogers, an Ainsworth friend of Marie's, was also a guest in camp on the same occasion.

I had met Francesca LaMont in Ainsworth a few evenings before when Marie had invited me to dinner. When Francesca joined us at the table, she was wearing a beautiful summer dress that she had obviously not bought at the local Ranchland or JC Penny store. Francesca was tall and dignified. She was not a classic beauty but her demeanor reflected a beauty of character that endures long after the superficial beauty of youth has disappeared. I had not often dined with such accomplished people as the Skinners, Ms. LaMont, and Mr. and Mrs. Rogers. For that reason, I was quite self-conscious about being there.

Marie was a gracious hostess who set a pleasant but country-casual table with delicious entrées that were often new to my palate. I was a meat and taters kind of guy. My mom was an expert at stretching a pound of hamburger into a meatloaf that would feed six. Mom tried her best to teach me basic manners but we didn't have a complete place setting of flatware. I had never had the opportunity to tackle a whole Rock Cornish hen by myself. That's what Marie served as the entrée that evening. At least that is what I remember it to be. Marie doesn't recall such an entrée. She may be right. I had not seen a Rock Cornish hen before so my identification may be as faulty as my early guesses at fossils that I found.

In Marie's defense, her dinners were never pretentiously formal. Nevertheless, from my limited experience, anything with cloth napkins was ostentatious. In addition to being all thumbs and painfully shy, my brain would usually clam up and refuse to offer anything to polite dinner conversation.

To the left and right of my plate that held a miniature chicken, I could see piece after piece of silverware laid out just for me. I am sure it was no more than a modest place setting made triple by my imagination. Marie confirms this. In those days though, more than one fork or spoon presented me with a dilemma—which to use first.

We had no formal dinnerware at home. We drank ice-tea from jelly glasses and were lucky if everyone at the table had a fork. My impression of formal and informal had been whether the artwork on the jelly glasses was of something refined like racehorses or game birds or something tasteless like Howdy Doody. Whenever I stabbed for a biscuit, my mom would warn me that I would someday regret ignoring her lessons in manners. That day had apparently arrived.

Did Mom say start at the edges and work in, or was it the other way around? I figured I could no more play that xylophone of flatware than I could play Beethoven on the piano. What I remember to have been a chicken looked like finger-food to me. Somehow, I just knew it wasn't. I waited and watched the others select their dissecting tools and I followed suit. I figured that I could never compete in etiquette at the level of Marie and Francesca. I just hoped to keep up with Morris.

There had been a few times, in my past, when I had run against the grain of Ainsworth aristocracy. Ainsworth is quite a friendly town, so those big fish in a little pond were very few in numbers. My class envy was sometimes due to their daughters' resistance to swimming with minnows like me. (The clerk in the five-and-dime is a case in point.)

Morris and Marie were, by any measure, high on the list of Ainsworth's most exceptional citizens. Nonetheless, they never placed themselves in any arbitrary class of exclusion. They befriended one and all. My first impression of Francesca LaMont was that she was from high and discriminating society.

To my surprise, Francesca LaMont was not like that at all. She quickly engaged me in conversation and actually listened to my answers. I could not believe that this refined, mature lady would be interested in my naive teenage point of view, but she was, and genuinely so. Before the dinner was over, I was no longer embarrassed about my simplistic answers to her questions. I believe it was the first time in my life that I realized that I had some worthwhile information stored up in my young brain. I didn't know everything but I knew some things. I had amassed a wealth of rural Nebraska lore that I had never before realized. Although Francesca was a world traveler, I was an expert in the unique things I knew. She casually and comfortably treated me as an equal at the table.

My association with Morris and Marie and their friends was always just that way. I have never met a more interesting and friendly set of people in my worldwide travels. Bone diggers are a friendly bunch, no question about it.

Until that dinner, I was apprehensive about Francesca LaMont's compatibility in a bone digging camp. My fears faded as I confidently played my flatware xylophone by ear. Delicious meal—refined company—nice memory.

Francesca adapted to her short stay in our bone digging camp with ease. Little did I know that this was not her first time in the field. Her days in a sleeping bag might have outnumbered mine. When nature called, she picked up the roll of toilet paper and shovel (We referred to those items as the "music roll" and the "wild-cat killer) and headed into the sagebrush like a veteran. She followed me around as I prospected and watched everything I did with intense curiosity. When it was time for camp chores, she watched how we did them once and pitched right in when they needed to be done again. Francesca LaMont was all right in my book!

The summers in Ainsworth presented a continual string of visitors to the Skinner home. Many times, we rendezvoused with others in the field. I met many paleontologists and geologists, nearly too numerous to mention by name. A quick scan of my diaries from 1961 and 1962, alone, shows the following names: Ted Galusha, N. Z. Ward, and Malcolm McKenna of the American Museum of Natural History; Walker Johnson and Richard Paull of Standard Oil; Reid McDonald from the University of Idaho (Moscow) and assistant Don Miller; Bob

Wilson from the South Dakota School of Mines; Al Agnew, South Dakota State Geologist, and his associates Bob Schoon and Tip Tipton; Harold Cook, the paleontologist and owner of Agate Springs Ranch south of Harrison, Nebraska (now Agate Fossil Beds National Monument); Ray Alf from Harvard University; John Clark; a Mr. Webb from California; Claude Hibbard and four assistants from the University of Michigan; a field crew from the University of Notre Dame; Ted White of the Dinosaur National Monument.

Friends without scientific credentials also visited Morris and Marie. I remember the artist George Mooy who went into the field with us and painted the Nebraska landscapes as we worked; Edmund Pease, an investment officer for the Chase Manhattan Bank; and Nurak Israsena, a Pharmaceutical Chemist for Charles Pfizer.

Nurak was from Thailand. It seems that he was umpteenth in line to the Thai throne. Nurak hosted a dinner at the Skinners and served a delicious curry that set our mouths on fire. I remember taking Ed Pease and Nurak to my dad's ranch on the Niobrara River. I enjoyed taking city-dudes out into the sticks; royalty was even better.

Nurak and Ed enjoyed shooting arrows from my hunting bow. My bow was quite powerful so if you weren't careful, an arrow with too much elevation could easily fly out too far to find again. I had lost so many arrows that I didn't think much about the ones that we lost that day. A few weeks after Nurak and Ed went back to New York City, I received a package from Abercrombie & Fitch, the renowned New York outfitters. There was only one Abercrombie & Fitch in those days. The package contained a half-dozen of the best fiberglass arrows and hunting tips, courtesy of Nurak Israsena. I still have them in my quiver in the attic. My archery was never sufficiently consistent to risk launching prizes like those into oblivion.

While I am on the subject of bows, I will tell you about a bow that hangs on the wall above our bed. It is an artifact that has unknown ultimate origins, but it is precious to me.

On June 17, 1966, Margaret Cook, widow of Harold, called Morris and urged us to come to her Agate Springs Ranch south of Harrison, Nebraska. There were complications in the plans of the new "Agate Fossil Beds National Monument." The planners, on-site, were not paleontologists and they could find no fossils in Carnegie and University Hills. Margaret worried that they would bring in heavy equipment and ruin what she knew to be a fossil-rich locality.

Morris and I arrived there the next evening and stayed with Margaret in the Agate Springs Ranch house. The house, built in the 1890s, is a historic spot in Nebraska. Various dignitaries such as the author J. Frank Dobie had used the bedroom where I slept. Many artifacts and relics of the Captain Cook and Red Cloud era were displayed around the house. The Cook library was filled with many books written by their early pioneer neighbor Mari Sandoz, as well as Charles Russell, Dobie and others.

On the morning of June 19, Morris and I went to the quarries and easily discovered the reason why the National Park employees could not find fossils. They were digging in the dump piles of earlier digs instead of the quarries—mystery solved. Jackson Moore, an archaeologist, was in charge of the project. He was experienced at excavating Indian mounds and was exceptionally careful and exacting about his work. Unfortunately, his profession had not prepared him for mining fossils. Once we had him turned around and facing the hill instead of the dump piles, fossils showed up in the first few scrapes of his trowel.

Before Morris and I departed the Agate Springs Ranch, Margaret served lunch beneath her cottonwoods. She had a Labrador retriever that loved to prove it. Margaret asked for my hat (a beat up old straw variety). She assured me that her dog would not harm it. She sailed it into the air like a Frisbee and the dog took chase. In its enthusiasm, it tried to retrieve my hat while it had a front foot firmly planted on the brim. A portion of the brim separated with a "r-r-r-rip" we could hear at the picnic table. I didn't mind that my hat now had a new window. It actually wasn't too bad being able to look at the sky and the ground at the same time. Margaret, however, was embarrassed.

When we were saying our good-byes, she gave me an old, wooden longbow. She said that it had been Harold's but she had no idea where it came from. It's easy to pretend that Red Cloud or Sitting Bull once owned it. I value it as a keepsake. I have hung a few feathers from it (all picked up in nature…wild turkey, red-tailed hawk, horned owl, etc.) and it pleases me as it hangs on the wall above our bed.

My memories of bone digging meals would not be complete without mentioning Marie's fantastic lunches she served on their long, screened-in porch. Every noon in Ainsworth, I would be in the packing-shed with my tummy growling in anticipation of Marie's delicious sandwiches of homemade bread stuffed with great piles of filling, lemonade and carrot-cake. What treats!

The packing shed was where we prepared our discoveries for ship-ment. We usually spent a week or so between expeditions in Ainsworth. We needed the time to build boxes, list and pack the fossils that we had found, and ship them to the museum. I can still remember taking a felt tipped pen and writing on each crate, "To: Frick Laboratory, American Museum of Natural History, Central Park West at 79th Street, New York City, New York." There was some pride in those pen strokes.

One time while in Ainsworth, Marie enlisted my help to rearrange her furniture. A misunderstanding could have resulted in something serious if it were not for the gentle, good humor of Marie.

One of Marie's prized possessions was an antique pump-organ. Morris and I moved furniture under Marie's direction, shuffling this one out and that one in, until we arrived at the proper combination. Another field assistant, Kenny Weichelman, joined our moving game quite late. We were down to one of the last few items, the pump-organ. Marie had assumed that we could move the remaining pieces without her help so she had gone about doing something else.

Morris said something and Kenny departed. In a moment, he came back with Marie. Marie asked, "Now what do you want Morris?"

Morris answered, "I don't want anything."

Marie replied, "You sent for me! Kenny came and got me."

Morris insisted, "I didn't send for you."

Marie declared firmly, "You did too!" She turned to Kenny and brought him into the discussion, "He did too! Didn't he Kenny? What did Morris tell you?"

Kenny quietly muttered, "He told me it was time to go get the old squawk box."

Morris pulled out all the stops, "The pump organ, Kenny—the PUMP ORGAN!"

I am not sure if Marie ever resolved how Kenny could jump to such a horrible conclusion. Perhaps she thought that Morris frequently referred to her in such terms when she was not around. Marie, rest easy. Morris never once spoke ill of you in my presence.

Chapter 21

Photo 17: From Kodak Point, South Dakota—Sep 1961

Is this insomnia
or am I absolutely normal?

The first night of our expedition would often find us at Kodak Point, south of Kadoka, South Dakota. Back then, the highway dropped abruptly over a rim of exposures. A short drive to the west of the high-way, across a buffalo grass meadow, would take us right to the brink of an extraordinary badland landscape. We didn't prospect there too often but it was a superb place for a stopover.

We would park the pickup into the wind, as the tailgate was our kitchen. The flames of the Coleman stoves were far more predictable if they were protected from the prairie breezes. If we detected mosquitoes, we would stick two long-handled shovels into the ground about fifteen feet away from and perpendicular to the front and rear of the pickup box. We tied ropes from the pickup to the two shovel handles. The ropes served as picket lines for both ends of our mosquito nets. We laid a tarpaulin on the ground under the square of ropes to serve as a floor to our bedroom. Our ceiling was the sky.

During the first few nights in camp, we would have fresh meat to fry in our skillets, and pots of boiled potatoes and canned vegetables to heat on the stoves. I was usually so hungry that I didn't mind the clean up chores. Scraping the meat cracklins from the skillet and sopping them, and the remaining grease, with pieces of bread was a delicacy I could not resist.

Another young colleague became quite scarce when domestic chores were waiting. He would wander into camp when dinner was ready. After he would eat, he would be off into the evening to pretend to log one more hour of prospecting, or he might take the "wildcat killer" off into the canyons and forget to come back until after dishes were done.

The teaching of camp culinary arts and etiquette is a process that takes forethought and skillful execution. You can't just scold, "Get your carcass over here and peel potatoes!" or "Don't forget dishes!" Such harshness could wound the tender egos of fledgling bone diggers. The new guy in camp could only be brought into compliance with skullduggery. There was no other way. Morris was a master and I was his willing accomplice.

Morris did not bring his gastronomic concoctions to perfection through a pinch of this or a pinch of that—a can would do. He would open whatever cans he could easily reach in the grub box and pour them all together into a big pot of what Chef Skinner called "slumgullion." Nearly every time it was different and every time we would wolf it down and wipe the pot clean with daubs of bread. We would then show our appreciation with smiles and reverberating belches followed by a chorus of F-sharps and flats. Morris would encourage our flattery by categorizing our responses or joining in.

When Morris and I went to town for grub, we dreamed up an exquisite new practical joke to teach this new guy a lesson. We noticed that corned-beef hash and dog food came in identical cans. We bought two cans of each. Before we put the cans into the grub-box, we switched labels. The corned-beef hash was now labeled "dog food." The real dog food went up into a sideboard bin where we would not serve it by mistake. Now it was just a matter of putting the plan into action.

At dinnertime, Morris prepared a slumgullion and left the two empty cans labeled "dog food" with the others. Our victim burped down his portion with gusto and commented, "Morris, this is even better than usual tonight. What's in it?"

Morris replied in an indifferent way, "Hell, I don't know. You know me. I just grab whatever is closest—a can of this and a can of that." Morris motioned to the tailgate of the pickup. "The cans are still over there."

Our new guy walked over and turned the cans to see the labels, "canned corn, green peas, dog food—*dog* food? It's *DOG* food!" He paced quietly and soon sat down on the grub box. He turned green-in-the-gills and made some preliminary motions to hint he wanted another look at the concoction he had eaten. Nevertheless, he fought the urge and kept it down.

Morris looked surprised. "Dog food? How the *hell* did that get in there?" Morris and I both agreed that the meal was slumgullion par excellence and deserved repeating some day.

From that day on, our new apprentice chef took a new interest in meal preparation. He was there each evening to see the ingredients of what he was about to eat.

A year or two before my employment, Carl Elfgrin, another young bone digger, was always hungry. Morris knew that it was a normal condition. When a young man walks from daylight to dark, he works up an appetite. Nevertheless, Morris, not to let a practical joke slip through his fingers, saw an opportunity. Morris mentioned, for a few days, that the hungry fellow might have a tapeworm. When Morris went into Casper, Wyoming for supplies, he found a willing druggist. He asked the druggist to fill the largest pill capsules he had with something totally harmless. The druggist found some capsules, probably meant for a horse, and filled them with a white powder. He then wrote an official prescription label for the bottle.

When Morris arrived back in camp, he gave the bottle of pills to the hungry assistant. Carl fought his tapeworm by choking down horsepills, morning and night, until he analyzed the label more closely. It read, "Lactase Purosa: For Worms." It didn't take a scholar to conclude that lactase purosa was pure milk sugar. Carl knew he had just enjoyed the distinct honor of being a target of Morris' practical jokes.

Morris also told of a time, many years before, when they were waiting out the heat of the day in an old schoolhouse or line shack. The windows were either non-existent or wide open for cross ventilation. One fellow, who thought himself to be quite the ladies' man, had fallen asleep in an open window. The others had grown tired of his continual stories of macho conquest. This appeared to be the opportune time for a lesson in humility. Morris went to the camp stove and heated some water to the perfect body temperature—so precise that he couldn't feel the water when he dipped his hand into it. He then proceeded to pour the tepid water into the sleeping man's crotch. They all sat back and waited. As the breeze cooled the water, the man aroused and looked down in horror. No one said a thing as the fellow slipped out of the open window and disappeared until the sun had dried his pants. Morris said that as far as he knew, the fellow went to the grave believing that he had wet his pants. Did it cure him of his macho boasting? Not likely!

Morris sometimes planned practical jokes weeks before he put them into practice. I remember walking along the shores of Alcova Reservoir, in Wyoming, with Morris and coming upon carp carcasses in various stages of decay. We noticed that a carp has a tooth that looks similar to a rodent molar. Morris carefully extracted a few. He later found a marmot jaw, traded teeth and presented it to a rodent-expert friend for identification. The rodent-expert puzzled over the Carpiomarmota for far too long.

Then there were the bull horn-weights. Some city slicker in the field noticed that bull's horns turned down and cow's horns turned up—an obvious sex-linked characteristic. Every time this fellow saw a bull in a pasture, he would comment about it. With his every comment came Morris' insistence that he should write a thesis on this previously unnoticed phenomenon. Therefore, the fellow kept field notes. It came time for the fellow to return to the city. Back in Ainsworth, Morris used bull horn-weights as candlesticks for centerpieces at his going-away dinner. Cattlemen clamp the horn-weights over a young bull's horns to pull the horn downward as it grows. The horn eventually arcs downward and inward at the tip. The in-turned points keep bulls from seriously injuring other animals. At the going-away dinner, the conversation turned to the heavy, lead candlesticks and Morris enthusiastically spilled the beans. They were what caused bull's horns to turn down! So much for a scientific thesis on sex-linked characteristics! The fellow returned to the city having painfully learned a secret of ranch life.

Practical jokes usually remained in camp. Whenever we headed into the fossil beds, we transformed ourselves into scientists. A fossil that had been waiting there for thirty million years didn't deserve a careless act that would render it worthless. A fossil can be collected undamaged, but it means nothing if it is not documented. The distinct layers of sediments are bookends for the fossils in between. When you empty your backpack after a day of prospecting, this little rabbit jaw might earn a title, "three feet below the purple ash." The label on that saber-toothed cat skull might read, "two feet above the nodule layer." Morris would make careful sections of each geologic incline and include every identifiable feature and the exact measure of elevation that separated them. Once the relationship of the purple ash to the nodule layer, for example, was established, the age of the fossils, as they related to each other, could be determined. Nevertheless, there were a few practical jokes, arranged out in the fossil beds, to teach lessons.

Photo 18: Oligocene Saber-toothed Cat Skull—Hoplophoneus
—July 1963

The grandson of Childs Frick, Townsend Burden III, spent a summer with us. Childs Frick was the philanthropist who made the Frick Laboratory and our employment possible. Townie was from a very rich family but you would never know it. He was like Teddy Roosevelt, an aristocrat who had a "bully good time" in the badlands. He was a big, friendly fellow who worked hard and thoroughly enjoyed the bone digging life. An uncommon failing was that, in spite of his carefulness, Townie sometimes set his backpack down and then would not be able to find it again.

"Thou shalt not leave your backpack!" was a bone digger's commandment. Therefore, whenever we came across Townie's pack we would add a rock or two. He would finally find his pack again and move on, only to leave it somewhere else. Another rock would mysteriously find its way into his pack. When he finally made it back to camp, he would discover that he had been toting a load of ballast. Townie slowly, but good-naturedly, learned to keep his backpack in sight.

In my seven years, I can't recall being the brunt of a practical joke. Morris may have had elaborate plans but I was too cautious to fall for them. Perhaps I fell for them one after the other and I just never knew.

When the evening meal was over and the chores were completed, the bone digging camp became a peaceful place for which I still yearn. Morris scribbled the day's events in his diary. During the years that I kept a diary, I followed his example. I would also turn on my small transistor radio and listen to whatever the AM dial would bring in. Stations then were few and far between.

Before Morris came into my life, I had hardly been out of Brown County Nebraska. I didn't know there was much beyond country music. Our radio at home, when we had one, was usually set to the "Grand Ol' Opry" or the "Louisiana Hayride" on Saturday nights. When my rebellious teenage years began, I sometimes sneaked the dial over to one of the stations that boomed out early rock and roll, all the way from the Mexican border, after the sun went down. Listening to Wolfman Jack was a pleasant teenage pastime in Ainsworth.

Morris had the latest technology, a tiny transistor radio that in addition to the AM band also received the new FM signals, a novel concept that was nearly static free. He fiddled with his tuner and found a classical music station out of Rapid City, South Dakota. It wasn't long before I turned my radio off and started listening to the peaceful strains of this new variety of music. How sublime it was to watch the sun disappear behind a craggy butte and watch the planets wink alive to the accompaniment of a tiny radio speaker's rendition of Mozart! I have never been to great concert halls, but they can't possibly be better than that.

If we were treated to a full moon, the night view was breathtaking. The badlands spires around us looked like the lunar landscape to me. Nights were our private times to relax and take stock of our surroundings.

Nighthawks would flit in the evening sky. Each erratic flit of their wings would be accompanied by their characteristic call "neep…neep" and each flit would bob them slightly higher. When they had reached their chosen zenith, they would fold their wings and plummet toward earth. At the last possible instant, they would break their dive with a startling roar. The growl of air through their wings, "borrrRRRRRrrrr," could make the uninitiated believe there was a lion in the bush.

Pronghorn antelope would sometimes watch us from afar. If we waved a dishtowel, their curiosity would sometimes lure them toward us. Sometimes we would hear a pronghorn whistle from quite nearby. By slowly scanning the horizon, we might find its characteristic curved horns, ears, and eyes—nothing more—peeking at us from over a ridgeline.

If we looked down, we might find an arrowhead that had been there for hundreds or thousands of years.

If we looked up, we could see the magnificent blanket of stars seeming to revolve around Polaris as the Earth turned beneath us. Sometimes we would be treated to a shower of meteorites or an occasional comet would appear on its never-ending journey. In the early sixties, something quite new also shared the night sky. Early artificial satellites, with names like Echo, drifted like tiny dots of light, on an unwavering path across the backdrop of stars. In about ninety minutes, if sleep had not found us, we might see the same satellite silently crossing the sky again, after it had completed an orbit of our little Earth.

During nighttime in a bone digging camp you could not help but question the meaning of life—of the universe. My bone digging camp is now my backyard in Texas. I have not grown tired of being under the stars. I still throw pebbles into the air and watch bats dip and catch them as I did from Kodak Point. The eternal "neep...neep" and "borrrRRRRRrrrr," of the Nighthawk are just as thrilling now as they were then.

Nighttime in a bone digging camp sometimes provides other treats.

On one midnight, there was a beautiful rainbow. The full moon cast an eerie arc of greenish-white light after a light shower had passed. Morris thought it was sufficiently rare and beautiful to wake me. I thought it was sufficiently rare and beautiful to wake Kenny. Kenny grumbled, "Damn it! I've seen rainbows before!" and went back to sleep. The night presents many rare and beautiful things that you will miss, if you choose to.

I awoke one night to the whiff—whiff—whiff of beating wings. I opened my eyes and discovered that a great horned owl was hovering about six feet above my bedroll. The owl was silhouetted against the moonlight in a beautiful but startling display. If we left our ground-tarps down for several days, field mice often built nests under them. I suppose the owl was hunting mice. I worried that my mop of dirty hair sticking from my bedroll might make a believable cottontail decoy, so I moved and the owl flew on. I often wonder what it would feel like to be aroused from deep sleep by a great horned owl trying to carry me off by my scalp.

On another night, I witnessed a much more startling exhibition. Our long days of walking and perspiring often resulted in nighttime leg cramps. Cramps regularly visited Morris and they were excruciatingly painful. Morris presumed we were depleting important minerals from our bodies through sweat. Therefore, he purchased a large bottle of salt tablets and put them in the sideboard of the pickup. If we thought about it, we would take a few during the day for prevention. Unfortunately, we usually sought them out only after the pains of cramps set in. On hot nights Morris enjoyed being as cool as possible. He never let underclothing or pajamas get between him and a good night's sleep. Except for a considerable amount of added weight, he slept exactly as he was born.

On that night, I awoke to a groaning and a commotion that was rather alarming. I peeked out of my bedroll right into the faces of two full moons, the one of green-cheese up high in the sky and another, much larger, hovering there about half as high as the owl had been. Morris was suffering from severe leg cramps. He had struggled out of his bedroll to get some salt tablets and then spilled them into the grass near the head of my bedroll. When I awoke, he was bent over, picking pills out of the buffalo grass by moonlight, and bellowing like a lonesome bull. That lunar sight eclipsed anything I had ever seen before—or since.

Anyone who has slept within earshot of Morris knows there was one other thing to contend with, his snoring. His sawing-logs were not a big distraction if he kept sawing. The annoyance that ruined my sleep was when he quit. Morris would buzz along for a while and then, in mid-breath, he would go silent. I don't mean silent like snoreless breathing. I mean silent! His silence would force me to hold my breath so I could listen more closely for his next sounds of life. I didn't look forward to dealing with a corpse before my morning coffee. When my lungs were threatening to implode and a split second before I exhorted,

"BREATHE MORRIS...BREATHE!" he would peacefully start sawing logs again. Aerobic exercise therefore went on night and day.

When I described setting up camp, I didn't mention setting up the tent. That's because we never set up the tent unless it was going to rain. Please don't infer that we set up the tent when rain was in the forecast. We seldom set it up even when the atmospherics prompted us into stanzas of the old cowboy song, "Well, it's cloudy in the west and it's lookin' like rain." We set up the tent only after a night of weathering rain, hail, and wind in the open and then only if the forecast was for more rain and the sky was still "cloudy in the west."[35]

Before I worked for Morris, I believed that a camping experience was not complete without a tent, a babbling brook, and a shade tree. We didn't often mess with the first, and the badlands were entirely void of the second and third. Low spots sometimes had puddles of water, though, and they were breeding grounds for squadrons of mosquitoes. For that reason, we camped high and dry. If you wanted to find our bone digging camp, you needed only to look for the highest mesa in the area. From it, if you could see an even higher place, that's where we would be. If it had rained, and the rain was threatening to repeat itself for the second night, that is also where we pitched the tent.

The tent had a metal frame that culminated in a spire about six inches long that held the center grommet. That spire pointed to the Heavens as if it defied the Almighty to put a lightning bolt right down it. The more physics courses I took at the university, the more I worried about the attraction of lightning to high, metallic places. It seemed risky to find the highest hill, construct a metal frame with a pointed tip, and sleep under it in a thunderstorm. If Ben Franklin lured in lightning with a key, I thought we were just asking for a science lesson.

---

[35] There were a few times in seven years when we set up the tent in advance of inclement weather. It was usually when we were in the northern latitudes, early or late in the year.

I had also discovered in my physics courses that air was a good insulator. Therefore, I would pump my air mattress as tight as a tick and would assume a rigid fetal position in its center to make sure I wasn't grounded. As the thunder boomed closer and closer, as the lightning popped like flashbulbs and as the wind whipped and flailed the canvas against the tent frame, my young mind presented me with endless questions:

- Will the metal tent frame conduct the lightning bolt into the ground and spare us?
- If so, will my eardrums survive being at ground zero?
- Does an air pump, that uses the compression of an internal combustion engine, truly pump fresh air?
- If not, is an air mattress full of gasoline vapors and oxygen safe around a spark the size of a lightning bolt?
- If not, can I mentally derive a differential equation that describes my path into orbit?
- Morris is running silent again; is he still among the living?
- Is this insomnia or am I absolutely normal?

Chapter 22

Observe the talus, Man!

Let's travel back to Kodak Point and my first evening as a bone digger. Although we were only camping overnight, Morris took me below the rim to teach me the first lesson of fossil hunting.

Imagine that it is 1960, that you are a sixteen-year-old kid from Ainsworth, Nebraska, and you are in the badlands with the renowned paleontologist, Morris F. Skinner, for the very first time. Your hometown is a little cow town, population of about two thousand, which proudly advertises itself as "The Middle of Nowhere." This morning you became one of the very fortunate, but utterly unknown, young men from the middle of nowhere, to be hired by Morris. It still hasn't quite soaked in that you, yes insignificant little you, are prospecting for fossils in the badlands and for the prestigious American Museum of Natural History. You remember your big brother telling you that Morris is no-nonsense when it comes to fossils. You are, of course, careful, tentative, observant, and hoping to do everything just right.

Imagine now that you are Morris F. Skinner who does not at all see his young helper as careful, tentative and observant. He sees the lad as an untrained retriever with boundless energy, a retriever who wants to bolt and chase every critter in the territory and flush every game bird beyond shotgun range. Morris sees a rambunctious lad who, when unleashed, will scramble off to mistake every fossil for a rock and whack it to pieces with his geology hammer just for the practice. Morris sees a novice who will spy a fossil some fifty-feet above him on a badland spire, and then scramble and claw his way up to claim it, no matter what havoc he creates along the way.

Here is what Morris told me, a sixteen-year-old, unknown kid from the middle of nowhere, on my first day. As we hiked into the badlands below Kodak Point, he told me not to act like a bull in a china shop but to take my time. He told me to be patient when I see fossils weathering out above me. He reminded me that those bones have been up there for millions of years; they will be there five minutes later. He warned me that if I scrambled and clawed my way up to the specimen, I would hopelessly bury important pieces in my mini-avalanche of talus. He told me that the first step is to check the talus for pieces that may have weathered out and tumbled down.

He then saw me return a blank stare, as I didn't know *talus* from my *tailbone*.

Morris explained that *talus* is all of the loose stuff that has washed down the slope to cover the underlying beds. If we start where the talus runs out at the bottom of the outcropping, and carefully scratch around, we will often find pieces of the fossil that is up there begging us to climb up and take a look. Therefore, we must take our time. It's like cleaning our plate and saving dessert for last.

He showed me how. What we later determined to be an oreodont skull was visible up the slope. An oreodont was a sheep-sized critter and a common animal in prehistoric times. We were prospecting in Oligocene sediments where oreodont skulls are a dime a dozen. At least they were in 1960.

Morris took a hook (a hand awl with a bend in the end) from his pack, and starting at the bottom of the talus slope, scratched like a mother hen looking for grain kernels. He would cluck his young chick over, point to things, and pick them up. By the time we had climbed to the oreodont skull, Morris had collected a handful of things. Not only did he have several pieces of the oreodont, but he had also found several small fossil rodent jaws, hackberry seeds, and a few coprolites.

He explained that a coprolite is fossil animal feces. The ones he held came from a carnivore, as bits of bones were visible within them. They looked like the landmines that neighbor dogs left around our lawn in Ainsworth, but these had turned to stone. Morris told me that carnivores often bury the remains of their kill and then defecate to mark the burial as theirs. He surmised that the oreodont might have met such an end and these coprolites are evidence of its demise. I was surprised to discover that my scientific job would entail collecting things that my mom could seldom get me to remove from the front yard for neither love nor money. Nonetheless, I was thoroughly impressed. Some of the coprolites I found that summer were never shipped to the Frick Laboratory. I couldn't return to my senior year in high school that fall without a few coprolites to show my friends. Coprolites are no Tyrannosaurus Rex, but they are impressive conversation pieces among teenage boys.

Once Morris had explained the coprolites, he showed me all of the other things in his hands. He showed me fossil rabbit and rodent jaws that look very much like the jaws of modern rabbits and squirrels. These, however, were over thirty million years old. He had fossil hackberry seeds that looked like those you can find beneath a hackberry tree, except these were also of stone. Morris held casts of prehistoric mud-dauber wasp nests and snails and other surprising things. These were all from the talus slope and we hadn't yet got to the prize—the oreodont skull that beckoned to us.

Except for the pieces of the oreodont, we could not determine the precise elevation from where all other specimens had eroded. They were interesting nonetheless. We could be sure that they came from above their resting places in the talus as gravity pulls everything downhill. How far above their resting place was anyone's guess. We isolated the pieces of the oreodont, placed them with the skull, and wrapped the other things in newspaper.

I discovered that there is a proper way to wrap a specimen. If it was fragile, we first wrapped a toilet paper cushion around it. We then laid it on a diagonal piece of newspaper, folded the near corner over it and rolled it over once or twice. We then folded one of the lateral corners across before continuing another half roll, and then folded the other lateral corner across. We then continued to roll the package and fold the lateral extensions into the center. This tapered the newspaper to the opposite point. When we had reached the end of the newspaper sheet, the specimen was effectively locked inside and had several layers protecting every dimension. Morris reached into his pack and pulled out a roll of brown-paper tape. He pulled off a length sufficient to encircle the package, while mustering up a mouthful of saliva. He stuck out his tongue, drew the full length of tape across it, and wrapped the tape around the package. He then wet the exterior of the tape with his tongue and wrote the field label onto it with an indelible pencil. The moisture activates the lead of an indelible pencil to produce purple ink. I still have my indelible pencil after all these years. It is sitting here on my desk, in a case made of two, brass .30-caliber shell casings that I found while prospecting. I scratched my name onto one and I stuck them over each end of the pencil. The cartridge casings kept my shirt pocket from turning purple when sweat activated the lead.

Morris labeled the package as precisely as possible, making sure that we had labeled it "talus" to show that the pieces inside had been displaced from their original locations.

Now that I have shared Morris' first lesson in fossil hunting, here is how you can put it into practice. Simply remember that impatience is not a virtue. Therefore, take time to enjoy the pursuit of your goals. By being observant, you will discover new things that bring meaning and fulfillment to your quest. When you take your time scratching through the talus of life for clues, you will realize that luck plays a very small part in your journey. You might use my little ditty to remind you of the only talisman you will need.

You hardly need a rabbit's foot or lucky amulet
or phylactery or bauble on which to place your bet.
A gewgaw or an icon or a magic ring won't do.
A periapt or fetish? You can do without those too.
To succeed in your pursuit of life, here's Step One of the plan.
Your number one priority is to observe the talus, Man!

Chapter 23

Photo 19: Preparing a titanothere skull and mandible for casting
—Seaman Hills, Wyoming—July 1960—Robert Emry (with shovel),
Alan Lamb and Morris Skinner (foreground)

I'm going to drink it all anyway!

Bits and pieces from the talus and other small durable specimens can be
wrapped in newspaper as I have described. Often, a specimen found in

a badland outcrop, such as a small skull of a horse, camel, oreodont or carnivore is in a nodule and sitting on a pedestal where erosion of the softer matrix has undercut it. Such fortunate finds are simple to collect. You can pluck them like cantaloupes, wrap them in newspaper and document your discovery.

When the specimen is as large as a rhino, titanothere or mastodon, then you are faced with a more daunting task. No amount of newspaper and wishful thinking will hold it all together. That's where the plaster of Paris and burlap are necessary.

Your complex chore begins with carefully trenching around the specimen. You can often use a pick and shovel on the extremes of the perimeter to hasten the task. The last few inches of trenching are done carefully and deliberately. If the pointed end of your geology hammer is sufficiently sharp, you can carefully chip away the clay. Then it's hook, awl and whiskbroom for the final work. You examine each piece of the surrounding substance as you remove it. If it shows bone, then you carefully put it back into place. When your trench is deeper than you presume the specimen to be, you then carefully undercut the block of material an inch or two on all sides.

You are now ready to pack out the large tools and pack in the plaster, burlap, water, rubber gloves, and plastic pan. If the block is quite large, you look for sticks along the way. The dead branches of a juniper or other woody resident of your locale will provide reinforcing for the plaster cast.

You first saturate any exposed fossil with shellac and then, after it dries, cover it with damp newspaper to keep the plaster from sticking to the bone. You then cut burlap bags into pieces with your knife. We each carried a knife with about a six-inch blade that we sharpened to a razor edge with a file. You then put on your rubber gloves, mix the plaster and water and began to encase the block. You slop pieces of burlap-drenched plaster completely over the top of the block, making sure that you work the plaster into any cracks and that you have several reinforcing layers extending into the undercut channel at its base. If your cast needs additional reinforcing, you hold the sticks in place with additional burlap and plaster.

When the plaster is still damp, you use your indelible pencil to draw an outline of where the fossils are located in the block. This is a good time to practice your artwork. If your rendition of a rhino skull looks more like a dragon, you might want to add some fire snorting from its nostrils. If there is room, you also write the geographic and geologic description of your find. Your documentation is no time to be creative; your details must be facts and not whimsy. Often there are associated pieces that are small enough to wrap in newspaper. You wrap them and make sure your documentation on the manila tape is complete and that all packages and casts are cross-referenced.

Now you wait for the cast to dry. You may sit in the shade, if you can find some, or you might eat your lunch or practice throwing your knife. If you can toss your knife by its blade and make it stick some twenty feet away, you try for twenty-five feet. If you are not alone, you can compete.

How long should you wait? You can't be impatient. Usually a properly cured cast will no longer be cool to the touch from the evaporation of moisture. I preferred to leave a bit of plaster in my plastic pan. When I could pop the leftover plaster from the pan and it remained in a single, discus-like piece, it was an indication that the plaster cast was also properly set. A too green cast will flex and lose its precious load as you try to turn it. If you are too impatient, the leftover plaster will glop out of your pan or crumble. You then won't have a discus that you can hurl over a badland ridgeline. If you wait for proper curing of your discus, the cast on the block of fossils will take care of itself.

When the plaster is dry and you have celebrated by hurling your discus over a ridgeline, you are now ready for the most nerve-wracking part of your chore—turning the cast. You usually take a few well-placed swings with the pointed end of your geology hammer at the base of the block. The idea is to break the whole block away from the matrix. When you have knocked it loose, you carefully turn it over. If you are convinced that the material will hold together, you can upend it with little trepidation. If the material that makes up the block is crumbly, you can use some help. Other hands can reach under the cast and hold things into place as you carefully lift it.

When the block is overturned, you sometimes see only clay. Your fossil is completely hidden. That is the optimum condition. If you have discovered that you didn't trench deeply enough, you are disheartened to see fossil extending into the pedestal where the block once sat. If this is the case, you carefully prepare another block, cast it and describe how the faces of both will integrate. This helps the preparators in the museum put it all back together.

Sometimes you have to make more than one block due to the specimen's size. A mastodon skull we collected in June Quarry, on Plum Creek near Ainsworth, Nebraska in the summer of 1963 was prepared in two blocks that totaled over four-hundred pounds. June Quarry was on the Eggers ranch. Charlie Eggers, who was a year ahead of me in high school, had built a rowboat in shop class. We enlisted the help of Charlie and his boat to float the large specimen up the creek to the Eggers barnyard. Charlie and Kenny Weichelman waded and pulled the boat while I pretended to push. It was the only time, during my employment, that we found a large specimen near enough to water to float it to civilization. It was also seldom that we found a large specimen within easy access of the pickup. Logistics was a continual problem.

If we succeeded in our goal of capturing the age-old critter in the first cast, and could see nothing but matrix on the bottom, we had to make an evaluation. This decision was based on how heavy the cast would be when the bottom was also encased in plaster, how far it was to the pickup, and how rugged the terrain. If we decided the load was within our rated horsepower, we would remove any large knots of obviously excess matrix, trim the edge of the plaster jacket, mix more plaster and enclose the bottom. Once it was dry, we would tote it out.

If Morris was handy and the block was large, he took on this task. Morris sometimes needed help getting to his feet after he had strapped a gargantuan block to his back. Nevertheless, once Morris was upright, with his powerful legs and low center of gravity, he could move a small mountain. He seemed to take pleasure in this challenge.

If I were the pack mule, I wouldn't be so confident. An ounce of material that I could otherwise leave behind would feel like a pound to me before I had climbed halfway out of the badlands. Therefore, before I plastered shut the bottom of the block, I would be careful not to disturb the specimen, but I would carefully remove every ounce of excess matrix. Even with modest loads, my backpack would nearly drag out my footprints as I plodded onward and upward.

As I trudged toward the pickup, I became convinced that this was a reason that Morris advised us to leave the drinking water there. The beckoning oasis was an incentive to keep homing in on it. When my sweat supply had dwindled to dust, I ignored the load on my back and worried about survival. *Survival of the Fittest* is a term common to evolution. I proved that *Survival of the Desperate* is also a workable theory that will drive your feet toward water.

Once we had arrived at the pickup and had dropped our heavy packs to the ground, we hastened to the nearest water jug. Those first frantic gulps of warm water were often accompanied by Morris admonishing us to take it easy—too much water too fast could make us sick.

My intense thirst and Morris' predictable comments reminded me of the story of a parched cowboy on a trail drive. When the herd had finally arrived at a watering hole, the cowboy jumped from his horse to guzzle muddy water from right beneath the cattle's feet. The other drovers went to the other side of the pond and asked the fellow why he didn't come over where the water was clean. He replied, "It makes no difference, I'm going to drink it all anyway!"

Chapter 24

Photo 20: Hand level once used by Morris Skinner and recently given to author by Marie Skinner.

Taking life one "eye-high" at a time.

When we had gained Morris' trust, he would let us carry a hand level and allow us to measure our own sections of geologic strata. He presented

me with a hand level that had been used by bone diggers who had preceded me.

My hand level had a worn leather case that hung from my belt. The
hand level itself resembled a telescope. It was a round, black, metal tube
with brass fittings that telescoped to about twice its length. To use it, I
would wrap my three middle fingers over the top and support it with
my thumb and pinkie. When I looked into the eyepiece, I saw crosshairs
and a bubble. When the bubble was centered, the crosshairs fell on
objects that were precisely level with my eye.

Morris told me to stand up straight and face him. He measured the
distance from my eye to the floor, making sure I was wearing my
prospecting boots. He said, "Your eye-high is sixty five inches. Don't
forget it." Technically, the height of a measuring instrument is abbrevi-
ated "H.I" for "height of instrument." In the case of a hand level, as
Morris used it, the "H.I" was also the height of the user's eye. "Eye-high"
is a more descriptive term and is easier to say, so "eye-high" is the term
we used.

The next few times we were in the field he would ask me to take out
my hand level. He would also take his out and we would measure the
same vertical exposures of sediments together.

We would select a starting point at the base—perhaps a volcanic ash
layer or a dividing line between two contrasting sedimentary layers. We
would stand straight and tall, with our boot soles on the starting point,
and sight to a point that was level with our eye. For me that was sixty-
five inches above the starting point, or one "eye-high." We would look
for something under our respective crosshairs that we would be able to
locate once we had walked there. It could be something quite distinct,
but most often it would be only a tiny pebble or an irregular feature in
the clay. Once we had settled on a feature, we would walk to that spot
and stand with our boot soles exactly on it. We would then stand tall
and take a second sighting through the hand level to another feature
that would be two "eye-highs" above our starting point. We continued

this process. All the while, we were sighting "eye-highs" and climbing, we would write descriptions of the underlying sedimentation in our notebooks. Once we completed our climb to the top of the exposed sediments, Morris would compare his notes to mine. I would multiply my number of "eye-highs" by sixty-five inches. He would multiply his number by the height that defined his "eye-high." Our products would be the vertical height of the sediments. If our products agreed and if our descriptions of the sediments matched, I had passed the test.

Morris had notebooks upon notebooks of such "sections" from all over the western United States. I do not believe it is exaggerating to state emphatically that Morris had completed the most comprehensive set of field records of any paleontologist or geologist, before or since.

I would only be bragging if I claimed that I contributed to his set of stratigraphy. He climbed every exposure and did every sketch, every calculation and every observation. I served him better by prospecting for fossils while he completed his sections. Sometimes though, I would follow along to back him up. I was smiling when I wrote "back him up." I could no more back up Morris, with a hand level, than I could play second fiddle to Isaac Stern.

Exact measurements are, of course, very important. There could be hundreds of thousands of years of sedimentation spanned in one "eye-high." A miscount or a careless sighting would make faulty all of the references we would later tie to these data points.

Accurate geologic sections from one area also assist in our determination of the age of sediments in a new area. If, for example, we can trace the same volcanic ash layer from butte to butte, we will expect to find the sedimentary layers to be in the same order and approximate vertical separation. If when we measure them with our hand levels we find differences, it could mean several interesting and very valid things. What we never ever wanted it to mean was that we had made a mistake. Life is too short and the badlands are too demanding to measure the same sediments twice. An axiom of carpentry is, "measure twice and saw once." This is not too practical for stratigraphy. Although I sometimes provided the second measure for Morris, he always made sure that my backup was not necessary. He was painstaking and so always trusted his figures over mine. Morris measured *once* with certainty.

Morris seldom missed an opportunity to check his own accuracy. One day we happened upon a crew from the U.S. Geological Survey measuring the elevation of a butte near Sheep Mountain, South Dakota. They, of course, had surveying transits on tripods. Their measurements would show up as contours on topographic maps. Therefore their measurements were even more time consuming and precise than ours. Morris asked the survey team where their benchmark was. A benchmark is a piece of pipe, or wooden stake, that they had driven into the ground to mark their starting place. Completed benchmarks often show the elevation and geographic location.

The surveyors pointed to their starting point and Morris and I headed up the butte with our hand levels. We reached the top long before the survey team. Morris' and my measurements were within a few feet of each other. When the survey team arrived, their official elevation measurement was very close to ours. We were encouraged to find how accurate we could be when there were so many opportunities to create small errors. The survey team, who had never been trained on hand levels, was also impressed with our quick results. The lesson that I learned was that every measurement device has a margin of error that you must accept. However, errors made through careless use of the equipment are unacceptable.

When I left for the Air Force in the fall of 1966, I turned in all of my fossil collecting equipment. Whenever I think of my bone digging days, I'm reminded of the hand level and what it taught me. The lesson extended beyond geology or stratigraphy. It taught me that if I selected suitable stepping-stones for myself, neither too low nor too high, then I would successfully achieve my goals. It also taught me to stand up straight and pursue things with confidence. A lack of confidence makes any mountain seem higher. Finally, it taught me that it is best not to inflate modest success through misrepresentations. If you crow from your highest mountain, others will look down on you.

Over the years, I have browsed antique malls and flea markets looking for a hand level like the one I used. I have little use for one now. I just wanted one to put on my desk and to fiddle with from time to time. Marie learned of my wish and graciously sent me a hand level that Morris had carried. It sits here on my desk beside my indelible pencil and picture of Morris and Marie. I'll show it to my grandchildren and tell them about my bone digging days. I'll tell them that a hand level is a tool for measuring life and I'll recite this bit of doggerel:

If you are feeling like the world has grabbed you by the tail
and you won't succeed without a life of crime,
then listen to this little verse. You can make it to the top.
You're not taking life one "eye-high" at a time.

If you are feeling sort of puny, like you just don't measure up,
and you shrink on down to a height that's just not you,
you will only see those shallow goals that fall within your crosshairs.
Your "eye-highs" to the top will be more too.

And when you start to tally just how high it is you've climbed,
those shallow goals will betray you in the end.
When you multiply their numbers by the height you truly are,
you will exaggerate the mountain you're on, my friend.

Or if you are feeling like a big shot and your present little stature
is not the height you think you ought to be,
don't cheat and call your "eye-high" exactly six foot four.
It won't change a thing your level lets you see.

You will climb your mountain wondering why it takes so many "eye-highs"
for a fellow who is the giant size of you.
When you multiply your "eye-highs" by the height you boast you are,
you will only think your mountain is bigger too.

Or if you just cannot take pleasure from that level's confining bubble
and you prefer to go just anywhere you please,
how will you know when you have reached your lifetime aspiration?
Will you ever see the forest for the trees?

My advice to you, if you are fed up where you are
and if you are also getting tired of my rhyme,
is to trust your old hand level. You can make it to the stars
by taking life one "eye-high" at a time.

Chapter 25

Photo 21: The author's hat on an Oligocene turtle—Oct 1961

I am a bone digger

When daybreak came, we would be off to our ultimate destinations, the fossil localities. This generic trip that I began with you found us at

Kodak Point. It could as easily have been many other places with descriptive geographic names.

My 35-mm camera captured many of the vistas that we visited. It seems that I left behind molecules of shoe leather and fibers from the knees of my jeans on every hill and dry wash. By the end of the collecting season, most of my boots had eroded away and my knees had long been showing through the denim. Except for a few pick marks here, and a ring of plaster over there where a fossil had been collected, the badlands showed little evidence of my passing.

I can't possibly take you to all of the places and describe all of the things we found. There are journals and lists that detail every little specimen—*Ictops* to oreodont—poebrothere to *Mesohippus*—titanothere to hackberry seed.

I hesitate to call those magnificent landscapes "badlands" because they were so very good to me. Respect them? Of course. Fear them? Never! The word "Badlands" is written on our maps to label our country's wild and desolate spaces. The place names, however, of the features that make the badlands what they are, reflect a broader understanding. The place names are so beautiful or so unusual that just listing some from my diaries seems appropriate. Some of the places I will mention were the backdrops to the fossil localities where we worked. Those vistas set the scene whenever we cared to look up, from millions of years in the past, to rest our eyes. The scattering of names, I believe, will bring back pleasant memories to any bone digger who had the pleasure of their association and may inspire others to seek them.

In North Dakota, there are the Fitterer Badlands, southwest of Dickinson, and Teddy Roosevelt's badlands near Medora.

In Montana, there are the Big Belt Mountains; the Lone Pine Hills; Ruby River; and Black Butte.

In Arizona, there are the Santa Catalina, Galiuro, Hualapai, and Aquarius Mountains; Redington; Sombrero Butte; San Pedro River; Wikieup; Milk Creek and Oak Creek Canyon.

In Wyoming there is Stinking Creek[36]; Bates Hole; Hat Creek; Spoon Buttes; Shawnee Butte; Wulf Ranch; La Prele Creek; Manville; Reno Ranch; Bed Tick Valley; Pumpkin Buttes; the Seaman Hills, Emerson Himes Ranch; Silver Ranch; Russell Thompson Ranch; Devil's Gap; Independence Rock; Alcova; Douglas; Lusk.

In Nebraska there is Old Woman in a Boat Butte; Bone Creek; Turtle Buttes; Devil's Gulch; Plum Creek; Niobrara River; Ash Hollow; Chimney Rock; Court House and Jail Rocks; Castle Rock; Roundhouse Rock; Scott's Bluff; Fort Robinson; Norden Bridge; Jones Canyon; Turkey Creek; Cub Creek; Chimney Creek; Deep Creek; Horse Thief Canyon; Agate Springs; Sand Draw; Rattlesnake Gulch; Harrison; Seven Sisters; Chadron; Trunk Butte; Snake River; Harris Ranch and on and on.

I especially like the place names from South Dakota. Many of them are translations of the Lakota or other Native American tongues such as Arrow Wound Table. Some describe their features such as Coffin Butte. Others are named after the early settlers such as Cuny Table. Jumbled together they sound like this: Pine Ridge; Porcupine Butte; Kodak Point; Grass Creek; Rocky Ford; Slim Buttes; the Castles of Reva Gap; the Big Badlands; Buffalo Gap; Flat Top Butte; Sheep Mountain; Babby Butte; Quiver Hill; Wounded Knee; Red Shirt Table; Rosebud; Windmill Butte; Cedar Butte; Oglala. The strangest to me is Squaw-Humper Table. My young mind often labored over how that desolate piece of real estate got its name.

My favorite place name is Plenty Star Table. Just say it out loud—"Plenty Star Table." Close your eyes and say it again—"Plenty Star Table." Isn't it beautiful? Doesn't it bring forth pleasant memories of your bone digging days? Can't you just see those millions of stars by

---

[36] It was listed as "Stinking Creek" on the topographic maps. Local ranchers called it by the four-letter word that begins with "S." We went up it many times without a paddle.

night? Can't you feel the intense bright sunlight of noon that even sends the shadows scurrying for cover? Can't you feel the soft evening breezes? Can't you hear the nighthawks? Don't you just know that "Plenty Star Table" means a lot to the Native Americans too?

So inspired, the bone diggers often named their quarries with strange and beautiful names: Devil's Jump Off Quarry; Horn Quarry; Swallow Quarry; June Quarry; Wade Quarry, Hollow Horn Bear Quarry; White Point Quarry; Observation Quarry; Running Water Quarry. Don't those names say something about the person who named them?

I believe that bone diggers are the last of the mountain men, the last of the fur trappers. If our era had had an Oregon Trail, we would have been on it. We would have dipped a gold-pan at Sutter's Mill. If we had not been a century, or two, too late, it would have been "Skinner and Emry" or "Skinner-Quinn" and not "Lewis and Clark" or "Warren-Snowden." Exploration is our life-blood. We are often driven by small things, but they are things that are very big to us. We know, beyond a doubt, that behind the next shovel of sand is the missing link that ties everything together.

I have not been a bone digger for nearly two-score years, yet I am a bone digger. Whenever I am out in nature, I snoop and poke around every dry wash I come upon. I am thrilled when I find a neighbor's cat skeleton near my back fence. I quickly find a shoebox and paintbrush and whisk around to find every possible fang and claw. The skull, sitting on top of all the rest, then stares out at me whenever I open my office closet door. What is the box of cat bones worth? Absolutely nothing. Why do I have them? I have them because I am a bone digger!

I will stick a hawk feather into my hatband and I will search for that perfectly smooth stone in the gravel of every stream. I will carry it as I walk and rub it with my thumb and I will thank Mother Nature for leaving it there just for me.

And less romantic, each time I wrap a package I am like Pavlov's dog. My mind subconsciously hearkens back to those times long ago when I wrapped specimens and found myself short of spit. I was often so dry that my tongue would clatter around my mouth searching for just a tiny bit of moisture to moisten the manila tape and to activate the lead in my indelible pencil. After forty years, I cannot wrap a package, even with scotch tape, without mustering up a mouthful of saliva. It is a conditioned reflex that certainly and scientifically proves that I am a bone digger.

Epilogue

Photo 22: Evening Primrose and Geology Hammer—Wyoming—July 1962

Years ago, I found myself daydreaming about those bone-digging days of my youth. I recalled the special times, especially my last weeks in the field with Morris. He and I were alone that glorious early fall of 1966. Others on the crew had returned to college or other endeavors and I was awaiting my late October orders to Air Force active duty. An early snow-fall drifted down one September night and we awoke to unmatched badland splendor. Being one-on-one with Morris was a privilege. Although he knew I was leaving for good, he taught me geology up to my final day.

Morris had not only mastered his profession of fossils and rocks, but he was a source of knowledge on all of nature. He taught me to see beauty and truth in common things and to sit in awe of majestic sunsets. I knew, during those weeks, that a unique but extremely important era of my life was ending, so I savored those days with some melancholy. I recognized that my days with Morris F. Skinner might well be the best of my life. Nonetheless, I knew that his positive influence would prepare me well for future endeavors.

Photo 23: Morris F. Skinner—Summer of 1963

Throughout the years of my Air Force career in fighter aviation, I did far-flung things that are only in most people's imaginings. In spite of this outwardly exciting life, I was unfulfilled. While stationed in Alaska, the Great Outdoors beckoned again. I became acquainted with Robert W. Service and his poetry of the adventurer. His *Call of the Wild* called to me. I thought of my Air Force medals and awards. I wondered why a simple Nebraska boy, who could have done far less with his life, should still feel discontented. I realized what I knew all along. My true interests weren't with high-flying and derring-do. Nature was always my calling—exploring it, learning from it, and writing about it.

My quick calculations, back then, proved that my mentor was well into his eighty-third year. I felt negligent that in the nearly quarter-century that had flown away in the Air Force, I had not let Morris and Marie know how much they meant to me. I wrote them the following letter:

January 13, 1989

Dear Morris and Marie,

It is going on 23 years since I laid down my geology pick in favor of stick and throttle. I have never regretted that decision; however, bone digging has remained in my blood and, now that "slipping the surly bonds of earth[37]" is behind me, memories of those earlier days become dearer and dearer to me with the passing of time.

---

[37] From John Gillespie McGee's High Flight:

"Oh, I have slipped the surly bonds of Earth
And danced the sky on laughter-silvered wings:
Sunward I've climbed, and joined the tumbling mirth
Of sun-split clouds—and done a hundred things
You have not dreamed of…"

Bone digging, to me, was not only the physical act of finding a fossil, collecting it, and sending it away to the museum, but it was a way of life, an education, a time to find myself. It was the fresh air, the sunsets, the antelope peeking over a ridgeline, the jokes we told, and the practical jokes that were the initiation rites of each new guy in camp. It was the knowledge of Mother Earth that we couldn't help learning when we were sleeping under the stars (in a sleeping bag tested on Mount Whitney) and examining each nook and cranny by day. It was the knowledge of the original Americans that comes from pondering over a mesa of teepee rings, a buffalo jump, or attempting the difficult task of creating a simple arrowhead. It was the few Lakota words I learned and familiarity with the Sioux traditions that we gained by working in their backyard. It was the studying of white man's history through our summer of following the Warren-Snowden route. It was climbing Chimney Rock to gather an ash sample, but also taking time to look out over the Platte valley and hear the ghosts of the wagon trains passing by. It was the attention to detail that each little specimen required if it was to become a meaningful advance to man's knowledge. It was the friendship and advice that you gave me. You made all of this possible through your leadership.

I would dearly love to find an excuse to tramp the canyons as we did back then. The summer we studied the Warren-Snowden papers was perhaps my most memorable. The hole-in-the-rock drawing that Snowden had completed some hundred years before and our finally pinpointing its location on the headwaters of the Niobrara—being able to walk left, right, fore, and aft and stand on the exact spot where Snowden did his sketch. To think about the Indian village that was in the forefront of Snowden's picture but had vanished without a trace. Flying a fighter is out of my system, but how I desperately wish I could be a "bone digger" on the trail of Snowden again.

I thought it was high time that I let you know how special you both are to me. In the words of Reader's Digest, "the most unforgettable character I have ever met" is Morris F. Skinner and of course his "better-half" Marie. I don't know if you are Robert W. Service fans, but he wrote my favorite poem. It reminds me so much of those bone-digging days. I'll include it for you.

Just in case you think I'm becoming a little too sentimental, too gushy or lofty in my praise of a gritty old character from my past, Morris, I also remember the meaning of your Indian name "Hosa Tonka", your "Classification of Farts" and that a proper measurement of thinness is a "gnat's ass stretched over a bass drum."

Sincerely,

Raleigh Emry

The Call of the Wild
Robert W. Service

Have you gazed on naked grandeur where there's nothing else to gaze on,
Set pieces and drop-curtain scenes galore,
Big mountains heaved to heaven, which the blinding sunsets blazon,
Black canyons where the rapids rip and roar?
Have you swept the visioned valley with the green stream streaking through it,
Searched the Vastness for a something you have lost?
Have you strung your soul to silence? Then for God's sake go and do it;
Hear the challenge, learn the lesson, pay the cost.

Have you wandered in the wilderness, the sage-brush desolation,
The bunch-grass levels where the cattle graze?
Have you whistled bits of rag-time at the end of all creation,
And learned to know the desert's little ways?
Have you camped upon the foothills, have you galloped o'er the ranges,
Have you roamed the arid sun-lands through and through?
Have you chummed up with the mesa? Do you know its moods and changes?
Then listen to the Wild—it's calling you.

Have you known the Great White Silence, not a snow-gemmed twig aquiver?
(Eternal truths that shame our soothing lies.)
Have you broken trail on snowshoes? mushed your huskies up the river,
Dared the unknown, led the way, and clutched the prize?
Have you marked the map's void spaces, mingled with the mongrel races,
Felt the savage strength of brute in every thew?
And though grim as hell the worst is, can you round it off with curses?
Then hearken to the Wild—it's wanting you.

Have you suffered, starved and triumphed, groveled down, yet grasped at glory,
Grown bigger in the bigness of the whole?
"Done things" just for the doing, letting babblers tell the story,
Seeing through the nice veneer the naked soul?
Have you seen God in His splendors, heard the text that nature renders?
(You'll never hear it in the family pew.)
The simple things, the true things, the silent men who do things—
Then listen to the Wild—it's calling you.

They have cradled you in custom, they have primed you with their preaching,
They have soaked you in convention through and through;
They have put you in a showcase; you're a credit to their teaching—
But can't you hear the wild?—It's calling you.
Let us probe the silent places, let us seek what luck betide us,
Let us journey to a lonely land I know.
There's a whisper on the night-wind, there's a star agleam to guide us,
And the Wild is calling, calling...let us go.

About the Author

Raleigh Emry is a native of north central Nebraska. He grew up on a ranch along the Niobrara River during the years immediately after World War II. It was a hardscrabble existence as the effects of the Great Depression and the War still had a hold on the land. After graduating from Ainsworth High School in 1961, he enrolled at Colorado State University and earned a Bachelor of Science degree in 1966. The Vietnam War and Selective Service draft were in progress. Therefore, upon graduation, he was commissioned as a Second Lieutenant in the United States Air Force. The majority of his over twenty years of military service was in fighter aviation including 218 combat missions in the F-4 Phantom II fighter while stationed in Vietnam.

His life long interests have been writing and exploring nature. It was only natural that, when he retired from active duty as a Major, he would pursue those interests. While Mr. Emry was in the Air Force, he had several articles published in official Air Force publications and won a USAF essay competition entitled "What Makes a Warrior?" Since retiring, he has had a short story of his boyhood days on the Niobrara published in NEBRASKAland Magazine. Works in progress are a novel about World War II and a book of poetry.

Raleigh Emry now lives in Manchaca, Texas, near Austin, with his wife, Terry, a 5th grade teacher.

0-595-22673-6